Nanotechnology

Nanotechnology

Molecular Speculations on Global Abundance

edited by BC Crandall

The MIT Press
Cambridge, Massachusetts
London, England

Fourth printing, 2000

For my father, my mother, and the dance of DNA.

This book was set in Helvetica and Melior by Graphic Composition, Inc.
Printed and bound in the United States of America.

Library of Congress Cataloging-in-Publication Data

Nanotechnology : molecular speculations on global abundance / edited
  by BC Crandall.
    p. cm.
  Includes bibliographical references and index.
  ISBN 0-262-03237-6 (hc : alk. paper).—ISBN 0-262-53137-2 (pb :
  alk. paper)
    1. Nanotechnology. 2. Molecular theory. I. Crandall, BC
T174.7.N375   1996
620—dc20   96–149
                                                        CIP

In all abundance there is lack.
—Hippocrates

■ Contents

This is an exercise in fictional science, or science fiction, if you like that better. Not for amusement: science fiction in the service of science. Or just science, if you agree that fiction is part of it, always was, and always will be as long as our brains are only minuscule fragments of the universe, much too small to hold all the facts of the world but not too idle to speculate about them.
—Valentino Braitenberg

We grasp the world at many levels and with a variety of instruments. Today, nanotechnology invites our species to grasp the world at an unprecedented level of granularity. The nascent sciences of molecular manipulation suggest that worlds of previously unimaginable material abundance are within our reach. For good or for ill, we stand at the threshold of a molecular dawn. How we conceive and create the instruments of nanotechnology will determine the quality and character of our lives in the next century and beyond.

Nanotechnology poses a difficult question: What will we human primates do when some of us learn to manipulate matter as finely as the DNA and RNA molecules that encode our own material structure? This book speculates on the outcome of this surprising and yet seemingly inevitable technological revolution.

Although inspired by a technical subject, this book is designed to be accessible to nontechnical readers. In the first chapter, I introduce the rudiments of molecular engineering and offer a brief genealogy of the field. I also provide a survey of several active areas of research. The remaining chapters present, with precision but without sustained mathematics, some of the potential power inherent in nanotechnology.

The book's first section focuses on applications that work, for the most part, inside living bodies.

In chapter 2, Ted Kaehler describes a molecular instrument for biological research, the "in-vivo nanoscope," that provides real-time images of living processes. Kaehler also discusses the possibility of entering the nanotechnological age almost instantaneously. If we can simulate molecular machines on existing computer systems, we may be able to download thousands of designs to the first molecular assemblers as soon as they are built.

Richard Crawford addresses, in chapter 3, the application of simple molecular machines to an immense and lucrative market: cosmetic surgery. Closely related is Dr. Edward M. Reifman's description in chapter 4 of what may be a very common application of early nanotechnology: diamond teeth.

The second section presents a wide range of applications that generally function outside human bodies.

In chapter 5, Harry Chesley presents a litany of every-day applications—from programmable paint to floorless elevators. Chesley also describes a simple molecular architecture that could support such applications. The focus of chapter 6, by John Papiewski, is a nanotechnological descendent of today's personal digital assistants, such as Apple's Newton.

In chapter 7, H. Keith Henson considers several whimsical applications of fairly advanced molecular engineering, including instantaneous cellular regeneration so that we can "enjoy" Conan-style blood sports—without actually dying. Tom McKendree, in chapter 8, concludes this section with a meditation on the enduring need for hobbies, even in an age of molecular machines.

The last section of the book considers extremely high resolution display surfaces and the functionality of generalized intelligent stuff.

In Chapter 9, Brian Wowk presents a description of next-generation flat-panel display devices and outlines the mechanisms of phased array optics. Finally, in chapter 10, J. Storrs Hall presents a detailed description of utility fog, which can act as a universal human-machine interface.

During the production of this book, two other books were published that deal specifically with molecular nanotechnology: *Nano,* by Ed Regis,[1] and *Prospects in Nanotechnology,* edited by Markus Krummenacker and James Lewis.[2] I recommend both books to those trying to understand this new field. Regis provides a journalistic account of the personalities involved, while Krummenacker and Lewis deliver the presentations and conversations of researchers at a conference held in 1992. I particularly enjoyed the fact that Regis used the title of a flyer I produced in 1991 as a chapter title in his book. Unfortunately he created a sound bite from my broadside that obscured its original intent. For the record, and because I believe its questions remain relevant, I reprint the flyer here, as a note.[3]

It is my pleasure to thank those who helped bring this book into existence. First of all, I am grateful to the writers for their contributions to this volume. I am also grateful to those who reviewed and commented upon all or part of this work, including Rick Danheiser, K. Eric Drexler, Jacqui Dunne, Gregory M. Fahy, Richard Feldmann, Dave Forrest, Michael Johnson, Ted Kaehler, Kirk Kolenbrander, Daniel T. Ling, Ralph Merkle, Hans Moravec, Charles Musgrave, Charles Poncé, Eric Raymond, Scott Roat, and Jeffrey Soreff. Responsibility for the text naturally devolves to myself and the other writers. I am also grateful to Terry Ehling of the MIT Press for her initial enthusiasm and continuing support of this project and to Sandra Minkkinen for her tenacious intelligence and good will during production.

I would also like to express my heartfelt appreciation to Susan Nance, Carol Meer, Samadhi Khan, and Pali Cooper for somatic recalibrations, and to Bibi Sillem for a place to call home with a view that just won't quit.

Mill Valley, California
May 1996

■ 1 Molecular Engineering

BC Crandall

A molecule is a collection of (one or more) atoms which are bound together by their mutual interactions for long enough to be observed as an entity. There is an enormous range in molecular stabilities and lifetimes, with some molecules existing only for the duration of an experiment lasting 10^{-12} s [a trillionth of a second] or less, while others may remain intact for billions of years. Molecules are sometimes classified by the number of atoms they contain (e.g., monatomic, diatomic, triatomic); those with three or more atoms are generally termed polyatomic. *Molecules range in size from monatomic species (as are found in gaseous helium or argon) to macroscopic aggregates (such as single crystals of diamond or quartz, and polymers), and in complexity from simple atoms to proteins, enzymes, and nucleic acids. Every known kind of atom is found in at least one diatomic (or larger) molecule; some atoms (e.g., carbon and hydrogen) are found as constituents of millions of different molecules. Molecules that contain carbon (with a few exceptions) are called* organic; *all others are* inorganic, *but these terms no longer imply a connection with living organisms.*
—Frank E. Harris

It is generally recognized that engineering is "the art or science of making practical application of the knowledge of pure science."
—Samuel C. Florman

☐ Nanotechnology Is Molecular Engineering

Nanotechnology is the art and science of building complex, practical devices with atomic precision. In an effort to construct useful machines, nanotechnologists apply the techniques of engineering to the knowledge generated by the sciences that study molecular structures. If we focus our attention on the developing capacity of these sciences in the closing years of the twentieth century, we can observe the birth of nanotechnology. In the next century, molecular engineering will emerge as a multitrillion dollar industry that will dominate the economic and ecological fabric of our lives.

This chapter begins with an overview of the scale of molecular objects. Then, following a brief introduction to atoms and molecules, and a cursory review of nanotechnology as an evolving discipline, the chapter highlights recent research efforts that are leading the dream of molecular engineering into reality.

- Scale
- Atoms
- Molecules
- A genealogy of nanotechnology
- Research frontiers

☐ Scale

To fully understand molecular engineering—to *be* a nanotechnologist—requires study in several fields, including physics, chemistry, molecular biology, and computer science. But this book is not a technical exposition of nanotechnology. Rather, it is an invitation to well-founded social dialogue as we collectively imagine and choose among different evolutionary paths. As such, the material presented does not focus on the analysis and mathematics required to actually design and build molecular machines. But we cannot escape the fact that it is precisely the scale at which nanotechnology operates that gives it its power. So, unless we intend to collapse into the simple repetition of superlatives, we need to come to terms with the dimensions of molecular objects.[1]

The goal of nanotechnology is the construction of a wide range of artifacts whose components are reasonably measured in nanometers, or *billionths* of a meter. At first it may seem impossible to imagine a billionth of a meter, or an apparatus with trillions of parts executing movements in quadrillionths of a second. But the human mind is a truly marvelous thing. With it we can imagine not only a frabjous day and a slain Jabberwock but also machines with precision-crafted molecular components. And while the former seem destined to remain figments, the latter images may someday be as real as you or me.

An imaginative exploration of nanotechnology requires that we conceptualize the two essential dimensions of any material system—space and time—as they relate to the molecular world.

First, how big are atoms? If, as the Greek philosopher Protagoras argued, "Man is the measure of all things," it is not surprising that we use a system of measurement with a unit of length roughly equal to the size of a human being. The metric system, which was proposed by Gabriel Mouton, Vicar of Lyons, in 1670 and adopted in France in 1795 subsequent to the French Revolution, is founded on two fundamental units: the meter and the kilogram. The meter is about forty inches long and a kilogram weighs a little over two pounds.[2] Because the metric system is a decimal system, all units can be expanded by multiples of ten. This allows for the use of a collection of prefixes, each of which can be affixed to a given unit, that indicate every third order of magnitude:

exa-	10^{18}	1,000,000,000,000,000,000	quintillion
peta-	10^{15}	1,000,000,000,000,000	quadrillion
tera-	10^{12}	1,000,000,000,000	trillion
giga-	10^{9}	1,000,000,000	billion
mega-	10^{6}	1,000,000	million
kilo-	10^{3}	1,000	thousand
	10^{0}	1	
milli-	10^{-3}	1/1,000	thousandth
micro-	10^{-6}	1/1,000,000	millionth
nano-	$\mathbf{10^{-9}}$	**1/1,000,000,000**	**billionth**
pico-	10^{-12}	1/1,000,000,000,000	trillionth
femto-	10^{-15}	1/1,000,000,000,000,000	quadrillionth
atto-	10^{-18}	1/1,000,000,000,000,000,000	quintillionth

The diameter of a single atom is a bit larger than one-tenth of a *nano-meter* (nm). Consequently, modest atomic constructions are most reasonably measured with these units. The DNA molecule, for example, is about 2.3 nanometers wide. Large, complex molecular structures may be measured in *micro*meters, or microns (μm).

Although most dimensions in this book are given in microns and nanometers, other scales are common in physics and biochemistry: the *angstrom,* the *dalton,* and the *atomic mass unit.* The angstrom (Å), named after the Swedish physicist Anders Jonas Ångström (1814–1874), is one tenth of a nanometer, or 10^{-10} meter. The dalton, named after the English chemist John Dalton (1766–1844), is a unit of mass nearly equivalent to that of a single hydrogen atom. The *kilodalton* (kd)—equal to a thousand daltons—is often used to indicate the molecular weight of moderate-to-large biomolecules. *Rho-dopsin,* for example, is a medium-sized structural protein that holds in place a small, 24-atom molecule called *retinal.* Retinal is responsible for the primary visual event in the rod cells of the retina. Rhodopsin weighs about 40 kilodaltons. That is, it has a mass equal to approximately 40,000 hydrogen atoms. The atomic mass unit (amu) is another unit that indicates the approximate mass of one hydrogen atom.

To get an idea of how small atoms are, Kenneth Ford offers the following image. "To arrive at the number of atoms in a cubic centimeter of water (a few drops), first cover the earth with airports, one against the other. Then go up a mile or so and build another solid layer of airports. Do this 100 million times. The last layer will have reached out to the sun and will contain some 10^{16} airports (ten million billion). The number of atoms in a few drops of water will be the number of airports filling up this substantial part of the solar system. If the airport construction rate were *one million* each second, the job could just have been finished in the known lifetime of the universe (something over ten billion years)."[3]

While atoms *are* incredibly small (see figure 1.1), we have, in the past decade, built the tools to not only "see" individual atoms but to manipulate them as well.

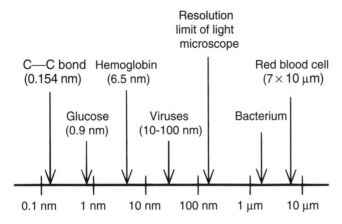

Figure 1.1 The relative size of atoms, biomolecules, bacteria, and cells. (After Stryer, *Biochemistry,* third edition, New York: W. H. Freeman and Company, 1988.)

The second dimension we need to imagine is the speed with which atoms move. In the atmosphere, oxygen, nitrogen, carbon dioxide, and other simple molecules zip around at up to ten times the speed of sound, or about 7,000 miles an hour.[4] But the local transformations that molecules are capable of are much more important than their gross physical movement as gaseous particles. For example, the primary event in vision—the mechanical transformation of the molecule retinal from bent to straight—occurs within 200 femtoseconds (*quadrillionths* of a second) after being struck by a photon.[5] Figure 1.2 shows the rates of a few biological processes.

In watery solutions, such as those found in the interior of most plants and animals, the general jostling caused by thermal excitation bumps protein-sized molecules into each other from all angles very rapidly. This bumbling, stumbling dance allows molecules to explore all possible "mating" configurations with the other molecules in their local environment. By variously constraining and controlling the chaos of such wild interactions, biological systems generate the event we call life. And it is exactly this mechanism of molecular *self-assembly* that may lead to the construction of complex structures designed by human engineers.[6]

One final note on scale. Although an atom is small, the nucleus is much much smaller, and of relatively little interest to nanotechnology. Kenneth Ford again: "To picture the nucleus, whose size is

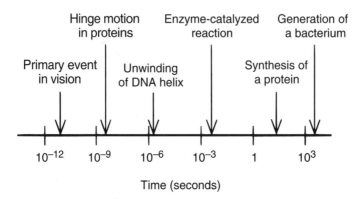

Figure 1.2 The relative rates of some biological processes. (After Stryer, *Biochemistry,* third edition, New York: W. H. Freeman and Company, 1988.)

about 10^{-4} to 10^{-5} of the size of an atom, one may imagine the atom expanded to, say, 10,000 feet (10^4 feet) or nearly two miles. This is about the length of a runway at a large air terminal such as New York International Airport. A fraction 10^{-4} of this is one foot, or about the diameter of a basketball. A fraction 10^{-5} is ten times smaller, or about the diameter of a golf ball. A golf ball in the middle of the New York International Airport is about as lonely as the proton at the center of a hydrogen atom. A basketball would correspond to a heavy nucleus such as uranium."[7]

The electron cloud that surrounds the nucleus determines the effective size of an atom, not the protons and neutrons at its core. Nanotechnology does not smash atoms to extract the energy held in the nucleus, as happens in nuclear power plants and atomic weapons. Nanotechnology is constructive; it snaps atoms together like Lego building blocks to build molecular structures in processes that are similar to, but potentially much more flexible and powerful than, the processes used by biological systems.

☐ **Atoms**

Atoms are forged in stars. Stellar fire consumes hydrogen atoms, the simplest and smallest element, and forges their nuclear components into heavier elements. Stars fill the universe with radiant light until their fuel is spent, then they collapse in upon themselves. If a star is large enough, this implosion becomes an explosion, and the collection of heavier-than-hydrogen atoms that was created by that star is thrown out into the universe.[8]

In time, a few of these atoms coalesce into planets, some of which may sponsor the material complexification we call life. On this planet, the Greek philosopher Democritus (460–370 BCE) was one of the first living creatures to suggest that the world consisted of very small but indivisible and indestructible pieces. These pieces were called "atoms" (*atomos,* indivisible: *a-,* not, + *temnein,* to cut). Democritus felt that everything that we see could be adequately explained as an aggregate or combination of atoms. And while "there are many different kinds of atoms, that is to say, they are not all the same size and shape . . ., no atom is large enough to be seen; conversely, no atom is so small as to have no dimensions at all."[9] Although we have moved from a philosophical to a scientific understanding of atoms, it is startling to realize that this ancient intuition of the underlying structure of matter is almost sufficient to comprehend the consequences, if not the precise machinations, of molecular engineering.

Human understanding of elemental structures evolved slowly until the late eighteenth and early nineteenth century when Antoine Lavoisier (1743–1794) proposed that elements retained their weight regardless of the compounds that they formed, and John Dalton (1766–1844) connected this concept with an atomic theory. Dalton is generally given credit for establishing the modern atomic theory with the publication of *A New System of Chemical Philosophy* in 1808, coincidentally the same year Ludwig van Beethoven first performed his Fifth Symphony. Dalton argued that the number of atoms in a gas, liquid, or solid, while numerous, must be finite, and that every atom of a particular element must be identical in nature. "His atoms were no longer smallest particles with some general and rather vague physical properties, but atoms endowed with the properties of chemical elements."[10]

Dalton wrote, "Whether the ultimate particles of a body of water, are all alike, that is of the same figure, weight, et cetera, is a question of some importance. From what is known, we have no reason to apprehend a diversity in these particulars: if it does exist in water it must equally exist in the elements constituting water, namely, hydrogen and oxygen. Now it is scarcely possible to conceive how the aggregates of dissimilar particles should be so uniformly the same. If some of the particles of water were heavier than others, if a parcel of a liquid on any occasion were constituted principally of these heavier particles, it must be supposed to effect the specific gravity of the mass, a circumstance not known. Similar observations may be made on other substances. Therefore we may conclude that *the ultimate particles of all homogeneous bodies are perfectly alike in weight, figure, et cetera.* In other words, every particle of water is like every other particle of water; every particle of hydrogen is like every other particle, et cetera."[11]

While Dalton's assumption that the simplest numerical combinations were in fact the actual ratios with which atoms combined led to some erroneous conclusions (e.g., that the smallest particle of water consisted of one oxygen and one hydrogen atom, rather than two hydrogens), his "new system" satisfied all known results and opened the door to rapid development of a modern atomic theory.

In passing, it is worth noting that the drawings Dalton used to present his theory in 1808 are flawed in that compounds are shown to consists of planar aggregates of circular atoms rather than three-dimensional collections of spheres. While atoms are clearly more complex than simple balls, the best first approximation of molecular form is that of closely packed spheres in three-space.[12]

Sixty years after Dalton's landmark publication, a total of sixty-three different elements had been effectively isolated and described in the emerging technical literature. The time was ripe for the imposition of order on this growing collection. Certain regularities were beginning to emerge and, in 1866, John Newlands attempted to articulate an overarching pattern. He earned himself the contempt and ridicule of the English Chemical Society for proposing that the elements could be meaningfully arranged by molecular weight in groups of eight—as notes in a scale—with every eighth element similar in many respects to the others so grouped. Newlands was right that there was a repetitive theme in the elements, but his structure was not sufficiently all encompassing, and his analogy to music was far too strange for his audience to appreciate. Newlands's career all but ended.

Three years later, as the first Atlantic-Pacific railway line was completed at Promontory Point, Utah, the Russian chemist Dmitri Mendeléev not only announced a cyclic pattern for organizing all known elements but he also accompanied his announcement with the declaration that three specific elements had yet to be discovered. As he constructed it, the "periodic" table (figure 1.3) contained several gaps, yet he was bold enough to predict not only the atomic weight but also the physical and chemical properties of several of the elements that would eventually fill out the table. As these elements

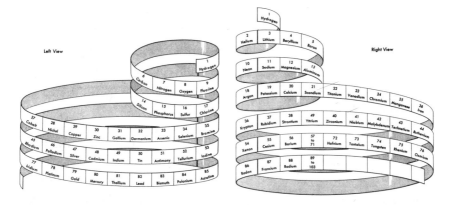

Figure 1.3 The Periodic Table. "The chemical elements have been arranged in sequence on a ribbon and coiled in a helix. As a result, elements with similar properties fall one under the other vertically. Two views of the coiled ribbon are shown so that one can see it from both sides at the same time." (From *Explaining the Atom,* by Selig Hecht, revised by Eugene Rabinowitch. Copyright © 1954, 1964 by Cecilia H. Hecht. Used with permission of Viking Penguin, a division of Penguin books USA Inc.) Hundreds of arrangements have been suggested for the presentation of the elements. For a sampling of three alternatives, and references to others, see Edward R. Tufte, *Envisioning Information* (Cheshire, Connecticut: Graphics Press, 1990, 14).

Figure 1.4 (*a*) Oxygen (O$_2$), (*b*) water (H$_2$O), and (*c*) ozone (O$_3$). (From *Molecules* by P. W. Atkins. Copyright © 1987 by P. W. Atkins. Used with permission of W. H. Freeman and Company.)

were discovered, Mendeléev's place in history was secured. The first missing element—gallium—was discovered four years later, in 1875, just as James Clerk Maxwell opened the next level of exploration by noting that atoms themselves were rather more complex than had been imagined.

"Having thus indicated a new mystery of Nature," wrote Mendeléev, "which does not yet yield to rational conception, the periodic law, together with the revelations of spectrum analysis, have contributed to again revive an old but remarkably long-lived hope—that of discovering, if not by experiment, at least by a mental effort, the *primary matter*—which had its genesis in the minds of the Grecian philosophers, and has been transmitted, together with many other ideas of the classic period, to the heirs of their civilization."[13]

☐ **Molecules**

Molecules are collections of atoms bound to one another. They comprise almost everything that we interact with. We breathe them, eat them, wash with them, and even think and feel with them. Solid matter contains on the order of 10^{23} molecules per cubic centimeter. Chemists estimate that there are 12 million specific molecular compounds on record, to which some 500,000 new compounds are added each year. This section introduces the barest few of this vast horde.[14]

As shown in figure 1.4, atmospheric oxygen is diatomic. Water, H$_2$O, is triatomic, consisting of two atoms of hydrogen and one of oxygen. Ozone is also triatomic, consisting of three oxygen atoms. When ultraviolet radiation from the sun strikes O$_2$ molecules (and other oxygen-containing molecules) in the upper atmosphere, it blasts the oxygen atoms apart. Single oxygen atoms then bond to an existing oxygen molecule (O_2) to form ozone. Ozone molecules are also torn apart by ultraviolet radiation. Each of these processes absorbs particular wavelengths in the ultraviolet portion of the electromagnetic spectrum. Bombarded by solar radiation, this frenzied activity proceeds continuously in a 10-kilometer layer between 25 and 35 kilometers above the Earth's surface. "The absorption of ultraviolet radiation by the gas is so efficient that at wavelengths near 250 nanometers, in the ultraviolet, only one part in 10^{30} of the incident solar radiation penetrates the ozone layer."[15]

Depletion of the ozone layer is dangerous for living things because the molecule that contains the instructions for building our bodies, DNA, responds dramatically to ultraviolet radiation. The DNA molecule has a natural harmonic at the frequency of certain ultraviolet light. When excited by this radiation, DNA vibrates excessively and falls apart like a bridge in an earthquake. Even though the DNA in healthy cells is constantly under repair, cancers such as melanomas can result when the disruption is too severe. In most cases, cancer is simply the result of a cell's best effort to follow a fouled set of instructions.

Methane, shown in figure 1.5, is a simple, flammable hydrocarbon made from five atoms. Ethane and octane, more complex hydrocarbons, consist of eight and twenty-six atoms, respectively. "The octane molecule results when we continue the process that leads from methane to ethane. Now a sufficient number of —CH_2— units have been introduced into the original C—H bond of methane to make the chain eight carbon atoms long (hence *oct*ane). Hydrocarbon molecules that contain half a dozen or so carbon atoms interact just strongly enough with each other to give a liquid at room temperature, so they are convenient to transport in tanks. But the liquid is still volatile and not too viscous to form a spray in the carburetor of an engine. Octane is representative of the size of the hydrocarbon molecules present in a gallon of gasoline. . . . Diesel fuel is less volatile: Its molecules are typically hydrocarbons with about sixteen

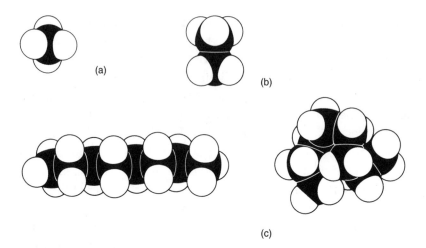

Figure 1.5 (*a*) Methane (CH$_4$), (*b*) ethane (C$_2$H$_6$), and (*c*) octane (C$_8$H$_{18}$) in two configurations. (From *Molecules* by Atkins. Copyright © 1987 by P. W. Atkins. Used with permission of W. H. Freeman and Company.)

carbon atoms in a chain. . . . Every hydrocarbon chain is actually a zigzag of carbon atoms and is flexible as well; moreover, each atom can be twisted around the bond joining it to its neighbor. A gallon of gasoline therefore contains some octane molecules that are rolled up into a tight ball, others are stretched out but still zigzag [see figure 1.5c], and others in the various intermediate configurations. Octane molecules are constantly writhing and twisting, rolling and unrolling, so that a gallon of gasoline is more like a can of molecular maggots than a box of short sticks."[16]

The double helix of DNA, which consists of thousands of adenine-thymine and guanine-cytosine pairs stacked one atop the other, is 2.3 nm wide. A single adenine-thymine pair is shown in figure 1.6 and eighteen contiguous pairs are shown in figure 1.7. Human DNA, which is found in almost every cell in the body, is nearly 10 meters long. To make the molecular components of the body, information is first copied from a strand of DNA onto a strand of *messenger RNA* (ribonucleic acid). This molecular chain leaves the nucleus and finds a *ribosome* in the cell where the information is used to assemble a string of amino acids or *peptides* into a *polypeptide*

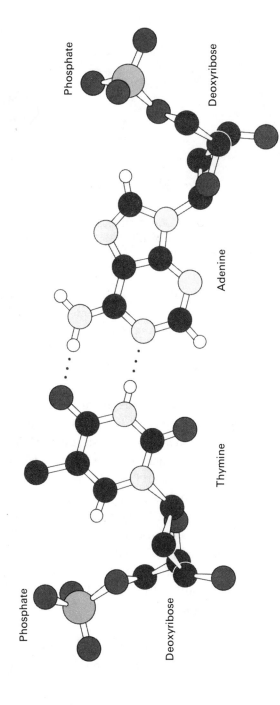

Figure 1.6 A single base pair of deoxyribonucleic acid or DNA. Adenine (26 atoms) is hydrogen bonded (dotted lines) to thymine (23 atoms) in this base pair. The other possible base pair consists of guanine and cytosine molecules. (From *Biochemistry*, third edition, by Stryer. Copyright © 1988 by Lubert Stryer. Used with permission of W. H. Freeman and Company.)

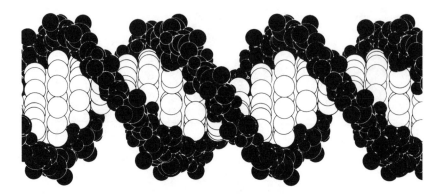

Figure 1.7 Eighteen base pairs of DNA, approximately 6 nm long and 2.3 nm wide. (From *Human Diversity* by Lewontin. Copyright © 1982 Scientific American Books. Used with permission of W. H. Freeman and Company.)

chain, which then folds in upon itself, forming a protein. Every living creature on earth utilizes this process to create its many components. (Some primitive viruses use only strands of RNA, without a DNA source.)

Ribosomes—roughly spherical structures with a mass of about 4200 kilodaltons in creatures more complex than bacteria—are the molecular factories for all living organisms. Ribosomes assemble proteins, which make up almost all living tissue, from a set of just twenty different amino acids. Reading a strand of messenger RNA, three "letters" at a time (see figure 1.8), a ribosome connects the appropriate free-floating amino acid (e.g., serine, alanine, valine, phenylalanine) to a growing polypeptide chain. Each amino acid is a simple molecule of fewer than thirty atoms; the simplest has only ten. It is the structural combination of these simple building blocks that creates the tremendous diversity of living organisms. Discovering the mechanisms that control the self-assembly of an amino acid chain into a biologically active, three-dimensional form is a major challenge facing molecular biologists today: the "protein-folding problem."[17]

Ribosomes are particularly interesting because they demonstrate that a simple molecular machine can, given the appropriate instruc-

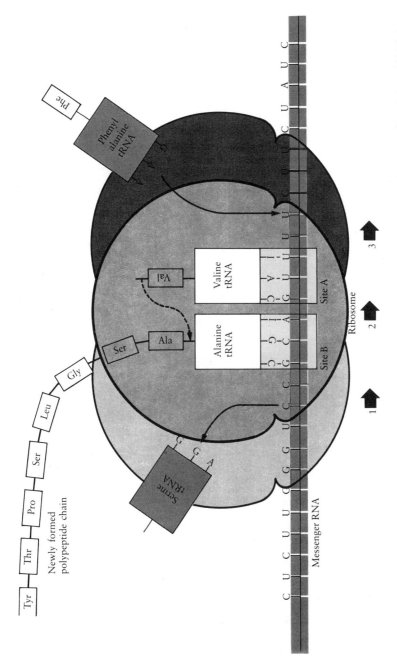

Figure 1.8 Ribosomal assembly of amino acids into a polypeptide chain. (From *Human Diversity* by Lewontin. Copyright © 1982 Scientific American Books. Used with permission of W. H. Freeman and Company.)

tions, assemble a vast array of other molecular structures. In fact, because ribosomes are themselves protein complexes, each ribosome is assembled by other ribosomes. From the perspective of nanotechnology, most of molecular biology, including the study of ribosomes, can be seen as a massive reverse-engineering project, where the machines in question are utterly without documentation.[18]

While ribosomes and other naturally evolved mechanisms of protein formation are today exploited by recombinant DNA technology and genetic engineering, the goal of molecular nanotechnology is the construction of *general-purpose assemblers* able to build molecular structures with any conceivable configuration—given the necessary instructions and assuming the structure is chemically stable. Assemblers with the capacity to build assemblers that are identical to themselves are called *replicators.*

Another biological system of particular interest to nanotechnologists is the molecular reaction used to power most changes in living cells. As shown in figure 1.9, the partial decomposition of adenosine triphosphate (ATP) to adenosine diphosphate (ADP) releases a very small amount of energy or heat. This energy is used to drive other chemical reactions that would not otherwise occur. For example, ribosomes consume a great many ATP molecules (turning them into ADP) as they coordinate the bonding reactions that assemble polypeptide chains. "The gearing of the reactions is such that the polypeptide is built at the expense of greater disorder elsewhere. To sustain these reactions, we need to eat, for we must rebuild the ATP from the ADP, using the energy of reactions stemming from respiration [for oxygen] and digestion [for sugars]. In a green leaf—one of the sources of the food that fuels us—the rebuilding of ATP from ADP is achieved by photosynthesis, and there the distant waterfall that drives the reaction is nuclear fusion in the sun. All life is the outgrowth of this intricately geared collapse into cosmic chaos."[19]

It is worth noting that humans have discovered and learned to use a molecule that is sufficiently similar to ATP that it can mimic its effects. Billions of us consume coffee, tea, and cola drinks because

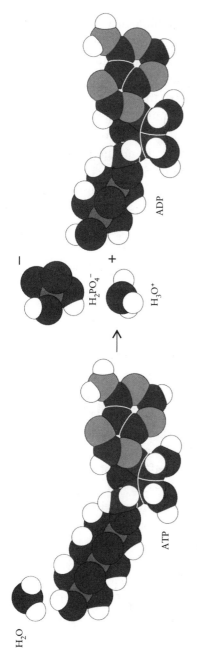

Figure 1.9 The reaction used to power many changes in biological cells, the transformation of adenosine triphosphate, or ATP ($C_{10}H_{11}O_{13}N_5P_3$), into adenosine diphosphate, or ADP ($C_{10}H_8O_{10}N_5P_2$). (From *Atoms, Electrons, and Change* by Atkins. Copyright © 1991 by P. W. Atkins. Used with permission of W. H. Freeman and Company.)

they temporarily increase our allotment of biological energy by providing our cells with caffeine molecules. Cocaine molecules, on the other hand, which used to be added to Coca Cola and other drinks, act like amphetamine. Rather than augmenting the supply of cellular ATP, amphetamine-like molecules insert themselves directly between neurons, mimicking the neurotransmitter norepinephrine.

On examination, all so-called drugs intervene in the processes of our bodies at the molecular level. A few simple molecules, such as nicotine, morphine, aspirin, cocaine, tetrahydrocannabinol (the active component of cannabis), and ethyl alcohol are responsible for the movement of billions of dollars a year—and none contain more than sixty atoms; most have fewer than thirty. While nanotechnology speaks to the possibility of a radically new level of molecular control, it is clear that we already live in a world of powerful molecules.

☐ A Genealogy of Nanotechnology

The companion fields of nanotechnology and artificial-life studies can be usefully thought of as being the *inside-out* of each other.[20] Each is dependent on and implicates the other; each is essentially useless and meaningless without the other. Nanotechnology addresses a collection of problems from a mechanical perspective, while artificial life—understood as the goal of an evolving effort in computer science to construct intelligent information systems—addresses essentially the same problems from an informational, systems-theoretic perspective.[21] Nanotechnology without artificial life can be little more than precise chemistry; artificial life without a nanotechnological embodiment offers little more than barren simulations.

Considering linguistic roots for a moment, we can see that "technology" derives from two Greek words, *tekhne,* meaning skill or art, and *logos,* meaning word or speech. "Nanotechnology" therefore indicates the discourse, hence the science, theory, or study of the skills required to craft matter at the nanometer scale. "Artificial" derives from the term "artifice," and comes to us from the Latin *artifex,* meaning craftsman: *ars* (stem *art-*), art + *-fex,* maker. "Life" is a much younger word, appearing in Old English as *lif* and seems to indicate the idea, "that which continues."[22]

A thorough genealogical treatment of nanotechnology and artificial life would therefore require tracing the evolution and development of not only the specific fields of chemistry, biology, engineering, and mathematics—especially applied mathematics as embodied in computer science—but it would also require an exploration of the anthropological roots of *homo faber,* the human artist, craftsman, and storyteller. Clearly, a brief overview must suffice.

If we allow that the primordial nanoscale "machine" is the double helix itself, we can begin by recalling a provocatively titled volume that appeared within months of our entry into the nuclear age. In 1945, on July 16, the United States detonated the first atomic weapon near Alamogordo, New Mexico. In less than a month, atomic weapons were exploded over Hiroshima and Nagasaki.[23] While the first bomb was in development, Erwin Schrödinger, a physicist who received the Nobel Prize in 1933 for discovering new forms of atomic energy, published a slim volume that asked a simple question: *What is Life?*[24]

Schrödinger's concern was not primarily philosophical. The question he asked was, "How can the events in space and time which take place within the spatial boundary of a living organism be accounted for by physics and chemistry?"[25] Schrödinger's essay is essentially an inquiry into the materiality of cellular life. "Adopting a conjecture already current, that a gene, or perhaps even an entire chromosome fiber, is a giant molecule, he takes the further step of supposing that its structure is that of an *aperiodic* solid or crystal. . . . Like other crystals, the chromosome can reproduce itself. But it also has another important attribute which is unique: its own complex structure (the energy state and configuration of its constituent atoms) forms a 'code-script' that determines the entire future development of the organism."[26] Schrödinger thus anticipated a key characteristic of the molecule of heredity—deoxyribonucleic acid or

DNA: its capacity to act as a set of instructions for the material construction of living forms.

In the same year, 1945, John von Neumann published his *First Draft of a Report on the EDVAC,* which described the basic concepts of the modern computer. Given the apparent similarities between organic systems and logical computation, it was inevitable that the mathematicians responsible for designing the first electronic computers would investigate the information-processing capabilities of both living creatures and engineered *automata.*

Von Neumann was an extraordinarily brilliant mathematician. In addition to consulting on the Manhattan Project and pioneering what would become the standard architecture for computing machinery, he "occupied himself with what he called *automata theory,* formulating axioms and proving theorems about assemblies of simple elements that might in an idealized way represent either possible circuits in man-made automata or patterns in organisms. Although he was stimulated by the problems of computers and organisms and continued to take account of them, he found the new abstract mathematics intrinsically interesting. One question with which he dealt in his automata theory is, If one has a collection of connected elements computing and transmitting information, i.e., an automaton, and each element is subject to malfunction, how can one arrange and organize the elements so that the overall output of the automaton is error free? The other problem was closer to genetics, . . . that of constructing formal models of automata capable of reproducing themselves. These would be models of a basic element of 'life.'"[27]

In 1953, as Tenzing Norgay and Edmund Hillary reached the summit of Mount Everest, James Watson and Francis Crick developed a convincing model for the "giant molecule of heredity," thereby validating Schrödinger's intuition. This new understanding sparked an explosion of investigations into the molecular mechanisms of organic life.

As biochemists busily studied the mechanisms of life as we know it, the physicist Richard Feynman pondered the physical limits of machinery as we might build it. In 1959, Feynman entertained an audience at the California Institute of Technology with a visionary presentation entitled, "There's Plenty of Room at the Bottom." In this talk, Feynman described a recursive process whereby very small machines could be built: each generation of machine tools would craft another generation with finer capabilities. At the conclusion of his talk, he predicted the arrival of atomically precise machinery. "I am not afraid to consider the final question as to whether, ultimately—in the very great future—we can arrange atoms the way we want; the very *atoms,* all the way down! (Within reason, of course; you can't put them so that they are chemically unstable, for example.)" [28]

In the years that followed, chemists and biologists focused on untangling the molecular structures that constitute materiality from the "bottom up," while physicists and electrical engineers devoted their efforts to building ever smaller machines from the "top down." Both efforts resulted in tremendously powerful technologies. Molecular science produced revolutionary medicines as well as the novel materials that provide the texture of so much of the modern world—including nylon, Tyvek, Teflon, and super glue—while micromachinists, after creating the first transistor in 1948, learned to build logic and computation machines with micron-scale components, generating thereby a global industry second only to agriculture.

The recent confluence of these two monumental efforts has produced an epochal cross-fertilization of knowledge—and the inevitable conceptual turbulence of two colliding world views. While top-down engineers such as computer chip designers build a fairly small number of complex machines as minutely as possible, chemists and other bottom-up technologists build relatively simple but atomically precise machines by the billions. Nanotechnology rises out of this confluence and aims at building complex, atomically precise machines by the trillions.

One of the most important developments for both approaches to the design and control of sub-micron systems was the invention of the *scanning tunneling microscope* (STM) in 1981 (figure 1.10). This

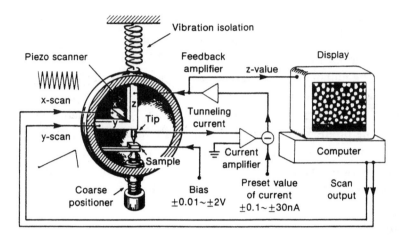

Figure 1.10 Scanning tunneling microscope. (From C. Julian Chen, *Introduction to Scanning Tunneling Microscopy,* Oxford: Oxford University Press, 1993.)

device, first developed by Gerd Binnig and Heinrich Rohrer at IBM's Zurich Research Labs, gave us the first direct images of individual atoms.[29] "The technique used ... involves an ultrafine stylus that hovers slightly above a conducting surface and senses topographic details via tiny fluctuations in the 'tunneling current' that forms between the stylus and the surface."[30] In effect, the STM senses the outer surface of the electron cloud that defines an atom. Figure 1.11 (Plate 1) displays a single atom defect in an otherwise perfect crystal lattice. The STM, which earned Binnig and Rohrer the 1986 Nobel Prize in physics, can image only conducting surfaces, but this limitation was overcome with the development of the *atomic force microscope* (AFM), in 1986, which can image nonconducting surfaces with similar resolution.[31]

Also in 1981, K. Eric Drexler, then a researcher at the Space Systems Laboratory of MIT, published a paper entitled, "Molecular engineering: An approach to the development of general capabilities for molecular manipulation," in which he argued that the natural mechanisms of protein synthesis demonstrate the feasability of human-engineered molecular machines. "To deny the feasability of advanced molecular machinery, one must apparently maintain either

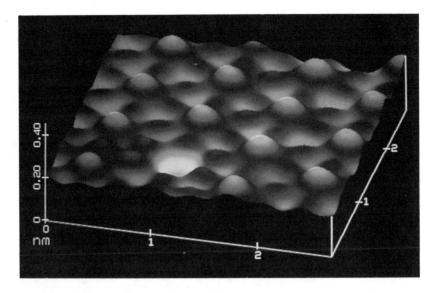

Figure 1.11 (Plate 1) STM image of a single atom defect in an iodine adsorbate lattice on a platinum surface, 2.8 nm square. (Courtesy Digital Instruments, Santa Barbara, California. Photo captured by a NanoScope® scanning tunneling microscope.)

(*i*) that design of proteins will remain infeasible indefinitely, or (*ii*) that complex machines cannot be made of proteins, or (*iii*) that protein machines cannot build second-generation machines."[32] Drexler argued that none of these objections can be sustained. In 1982, Drexler introduced the concept of molecular engineering to a popular audience with the publication of "When molecules will do the work" in *Smithsonian* magazine.[33]

In 1985, the discovery of a new form of carbon molecule stunned the scientific community. Carbon participates in a huge variety of molecules, but only two pure-carbon forms were previously known: graphite, which consists of two-dimensional sheets, and diamond, which is a three-dimensional network of interlinked atoms. In contrast, the molecule *buckminsterfullerene* (figure 1.12) contains exactly sixty carbon atoms in the shape of a soccer ball. In 1991, *buckyballs,* as they came to be known, were heralded as the "Molecule of the Year" by the American Association for the Advancement

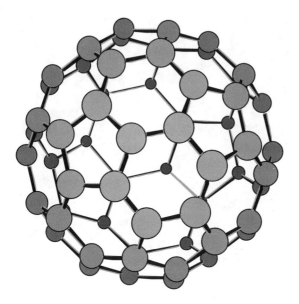

Figure 1.12   Buckminsterfullerene (C_{60}). A third form of pure carbon.

of Science and appeared on the cover of *Scientific American.* Named after R. Buckminster Fuller, these roughly spherical molecules have been nominated for a number of diverse applications, and their structure has led many to imagine them as viable components in various nanotechnological efforts.[34]

In 1986, while Binnig and Rohrer received their award in Stockholm for revealing to the world images of individual atoms, K. Eric Drexler published the first full-length book on the potential of molecular engineering for a popular science audience. *The Engines of Creation: Challenges and Choices of the Last Technological Revolution* introduced most of the writers in this volume, as well as many others, to the field of nanotechnology.[35]

In the following year, 1987, the first workshop on artificial life was held in Los Alamos, New Mexico. Jointly sponsored by the Center for Nonlinear Studies at the Los Alamos National Laboratory, the Santa Fe Institute, and Apple Computer, Inc., "the workshop brought together 160 computer scientists, biologists, physicists, anthropologists, and other assorted '-ists,' all of whom shared a common interest in the simulation and synthesis of living systems."[36] Several papers presented at this ground-breaking conference demonstrate the confluence of top-down and bottom-up approaches to molecular systems design, including a survey article by Conrad Schneiker, "NanoTechnology with Feynman Machines: Scanning Tunneling Engineering and Artificial Life." Schneiker suggests that "there are many ways that nanotechnology can eventually be applied to the development of artificial life. . . . (1) We can start with a completely natural life form and gradually transform it (bootstrap it) into a totally artificial life form by using molecule-by-molecule replacement. (2) We can develop a hybrid living system that incorporates some nanotechnology for computing functions, and some microtechnology for artificial replication."[37]

In 1988, Richard Feldmann, a computer scientist at the National Institute of Health (NIH), presented a paper, "Applying Engineering Principles to the Design of a Cellular Biology," in which he argued that building a biology is a reasonable goal for the scientific community. Feldmann voices an optimism familiar to computer scientists: "With exponentially increasing computer power, it will take far less than 80 years to be able to design and implement a biological system. The issue seems to be simply one of deciding we want to do a project like this, not the technological complexity of the project *per se*."[38]

In the same year, Hans Moravec, the Director of the Mobile Robot Laboratory at Carnegie Mellon University, published *Mind Children: The Future of Robot and Human Intelligence*. In this book he describes "a nanocomputer [that] might have a processing speed of a trillion operations per second. With millions of processors crammed onto a thumbnail-size chip, my human-equivalence criterion [for intelligence] would be bested more than a millionfold!"[39]

Two years after the first conference on artificial life, the first international conference on nanotechnology was held in Palo Alto, California, sponsored by the Foresight Institute (founded by Drexler) and the Global Business Network, and hosted by the Department of Computer Science at Stanford University. The volume that resulted from that conference, which I had the pleasure of editing, presents a broad range of technologies that contribute to nanotechnology as well as several perspectives on the consequences of success.[40] "Here," as one reviewer wrote, "in a highly accessible format, is a discussion of atomic probe microscopes, self-assembly in molecular crystals, protein engineering, micromachining and much else."[41]

In July of 1990, the English Institute of Physics launched a new journal called *Nanotechnology*. The ground-breaking issue included articles on various submicron technologies, including, "The Scanning Tunneling Microscope as a Tool for Nanofabrication."[42] Also in 1990, IBM set a record for miniaturized publicity—and brought nanotechnology to the attention of the popular press—by spelling out their company logo with thirty-five xenon atoms on a nickel crystal.[43]

In 1991, several companies announced their intentions to invest in nano-scale research. IBM's Vice President for Science and Technology, J. A. Armstrong, wrote, "I believe that nanoscience and nanotechnology will be central to the next epoch of the information age, and will be as revolutionary as science and technology at the micron scale have been since the early '70s. . . . Not only do we have the ability to make such nanostructures, but, as an outgrowth of the invention of scanning tunneling microscopy, we have the micromechanical ability to manipulate, as well as to see and measure, these structures. . . . Indeed, we will have the ability to make electronic and mechanical devices atom-by-atom when that is appropriate to the job at hand."[44] In Japan, the Ministry of International Trade and Industry (MITI) also announced the funding of a broad nanotechnology research effort.[45]

Public awareness of molecular engineering increased in 1991 as the American Association for the Advancement of Science devoted a special issue of their journal *Science* to the field, with articles on reverse engineering biological systems, molecular self-assembly, atomic and molecular manipulation, and investments in the fledgling technology in the United States and Japan.[46] In the same year, the *New York Times* reported on the second United States conference on molecular nanotechnology. Andrew Pollack, science editor for the *Times,* wrote that, "The ability to manipulate matter by its most basic components—molecule by molecule and even atom by atom—while now very crude, might one day allow people to build almost unimaginably small electric circuits and machines, producing, for example, a supercomputer invisible to the naked eye. Some futurists even imagine building tiny robots that could travel through the body performing surgery on damaged cells."[47] Exploring a range of such possibilities, Michael Flynn published a work of science fiction in 1991, entitled *The Nanotech Chronicles.*[48]

In 1992, Drexler, who is perhaps most responsible for promulgating a vision of molecular robots "performing surgery on damaged cells," moved to substantiate his musings with publication of *Nanosystems: Molecular Machinery, Manufacturing, and Computing.*[49] In this technical work, Drexler presents several mechanisms—bearings, gears, cams, clutches, as well as computational elements—designed to provide the basic components for nanotechnological assemblers. Though we cannot build these structures today, Drexler argues that assemblers will be feasible within the next few decades. While Drexler's work has received mixed reviews, his effort to describe the details of nanotechnological machinery is currently the most thorough articulation of molecular engineering's potential.[50]

Drexler is convinced that molecular assemblers will be able to build large-scale, molecularly precise structures very rapidly (one-kilogram objects in under an hour) and power them with billions of microscopic computers (each smaller than a blood cell) capable of generating more than 10^{16} (10 quadrillion, or 10 million billion) operations per second.[51] For comparison, the current goal for "high-performance computing" in the United States and elsewhere is the construction of a *teraflop* machine, which would generate 10^{12} (one trillion) operations per second, whereas scientific workstations of the mid 1990s are reaching to perform 10^9 (one billion) operations per second and most desktop computers perform no more than 10^7 (10 million).[52]

Whether Drexler's particular vision of the future comes to pass
(see figure 1.13), the continuing research efforts of thousands of molecular scientists and technologists—regardless of their chosen
title—seem destined to produce an exponentially increasing capacity to build molecularly precise structures, and this capacity will
lead to the construction of ever smaller and ever more powerful
computational and robotic systems.

In 1992, in order to support the efforts of a variety of molecular
researchers, the British journal *Nature* held their first nanotechnology conference in Tokyo. IBM's Don Eigler (the man who spelled
out their logo with xenon atoms), Richard Smalley (codiscoverer of
buckyballs), and others made presentations.[53]

The proceedings of the second conference on artificial life, held
in Santa Fe in 1990, also appeared in 1992,[54] as did two introductions to the field of artificial life: M. Mitchell Waldrop's *Complexity:
The Emerging Science at the Edge of Order and Chaos* and Steven
Levy's *Artificial Life: The Quest for a New Creation*.[55]

In the conference volume, J. Doyne Farmer, a physicist in the Complex Systems Group at the Los Alamos National Laboratory, and his
wife, Alletta Belin, write that "Within fifty to a hundred years, a
new class of organisms is likely to emerge. These organisms will be
artificial in the sense that they will originally be designed by humans. However, they will reproduce, and will 'evolve' into something other than their original form; they will be 'alive' under any
reasonable definition of the word. These organisms will evolve in a
fundamentally different manner than contemporary biological organisms, since their reproduction will be under at least partial conscious control, giving it a Lamarckian component. The pace of
evolutionary change consequently will be extremely rapid. The advent of artificial life will be the most significant historical event
since the emergence of human beings. The impact on humanity and
the biosphere could be enormous, larger than the industrial revolution, nuclear weapons, or environmental pollution. We must take
steps now to shape the emergence of artificial organisms; they have
potential to be either the ugliest terrestrial disaster, or the most beautiful creation of humanity."[56]

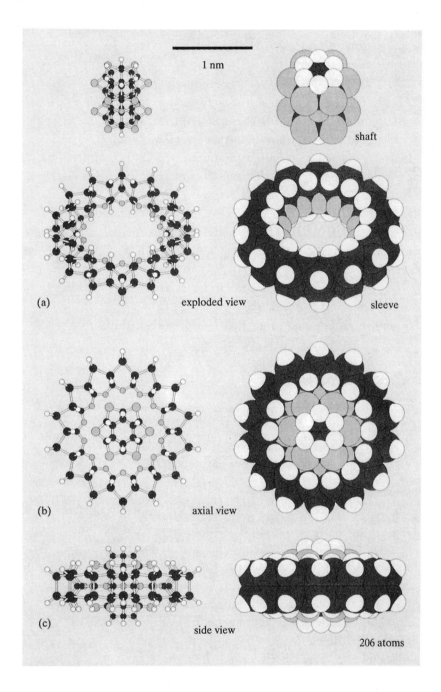

1 nm

shaft

(a)   exploded view   sleeve

(b)   axial view

(c)   side view

206 atoms

Figure 1.13 Three views of an "overlap-repulsion" molecular bearing. (From K. Eric Drexler, *Nanosystems: Molecular Machinery, Manufacturing, and Computing.* Copyright 1992 John Wiley & Sons. Figure courtesy K. Eric Drexler.)

Also in 1992, the Institute for Scientific Information noted that the prefix "nano-" was one of the most popular among new journals, including *Nanobiology* and *Nanostructured Materials*.[57]

In 1993, the National Science Foundation (NSF) committed funding for the formation of a National Nanofabrication Users Network, to include Cornell University, Howard University, Pennsylvania State University, Stanford University, and the University of California at Santa Barbara. The NSF's announcement noted that "Nanofabrication is a critical 'enabling' technology for a wide variety of disciplines. . . . The network will help the nation remain at the forefront of many burgeoning research areas, a number of which have commercial applications."[58]

Sooner than Feynman imagined—in less than thirty-five years— "the very great future" is at our door.

☐ Research Frontiers

Shifting from the chronicles of a genealogy to the necessarily chaotic impressions of current events, we conclude with some snapshots of recent research.

Atomic and Molecular Sensors

Tools that allow us to observe phenomena allow us to construct tools of manipulation as well as more refined tools of observation. Several sensing and imaging technologies have been developed that provide atomic and molecular resolution. Chief among these are the scanning probe microscopes, including scanning tunneling microscopes (STM) and atomic force microscopes (AFM). "In the 10 years since it first showed up in the laboratory, the scanning tunneling microscope has distinguished itself as a workhorse in the scientific instrument stable."[59] Priced between $50,000 and $500,000, and quickly becoming a $100 million industry, scanning probe microscopes provide an immediate window on the atomic world.[60]

Picosecond time-resolution is being added to the nanometer spatial-resolution of the STM. Researchers at IBM's Thomas J. Watson Research Center report that the "ability to combine the spatial resolution of tunneling microscopy with the time resolution of ultrafast optics yields a powerful tool for the investigation of dynamic phenomena on the atomic scale."[61] This work extends the imaging capacity of the STM from three dimensions to four, allowing observation of molecular interactions in appropriate length and time scales.

The STM's imaging capacity is also being extended to more accurately resolve molecular structures. Because the STM responds to the electron cloud surrounding an atom, and because atoms bound in molecules share electrons in complex ways, it remains difficult to identify molecular species with scanning probe microscopes. However, recent work at IBM's Almaden Research Center on STM image-processing algorithms has made it possible to image molecular structures with much improved clarity. Their work has "brought researchers a step closer to using STM to track the reactants, intermediates, and products of chemical reactions."[62]

Graphically demonstrating Louis de Broglie's 1942 theory that all elementary particles behave as both waves and particles, researchers at IBM's Almaden Research Center used an STM to create and image a circular "quantum corral" of forty-eight iron atoms on the surface of a copper crystal. The corral, fourteen nanometers in diameter, induced the copper electrons to dramatically display their wavelike nature by exciting the otherwise planar array of copper electrons into a standing wave pattern (figure 1.14, plate 2). "When electrons are confined to length-scales approaching the de Broglie wavelength, their behavior is dominated by quantum mechanical effects."[63] In short, the corralled atoms share their electrons to create a standing wave of quantum-mechanical densities that the STM can perceive. The STM was also used to nudge the iron atoms into place.

Buckytubes and Other Nanotubes
Elongated tubes of carbon, discovered in 1991 as a byproduct of buckyballs, became known, naturally enough, as *buckytubes* (figure 1.15). The tubes are roughly the diameter of buckyballs, but they

Figure 1.14 (Plate 2) STM of 48 iron atoms in a circle (14.3 nm in diameter) on a copper crystal surface demonstrates the wave-particle duality of electrons. (Courtesy IBM Corporation, Research Division, Almaden Research Center.)

can be several microns in length. Although reliable manufacturing techniques have yet to emerge, the structural properties are known to be unprecedented. Richard Smalley claims that "they would be the strongest fibers we could make from anything."[64] Discovered by Sumio Iijima, a physicist at the NEC Fundamental Research Laboratory in Japan, the carbon tubes have already been recommended for many applications. For example, "two concentric buckytubes may form a wire with an insulating outer shell and a conducting inner shell. This is possible because a buckytube's electronic properties vary with its diameter and other dimensions. The difference is not chemical—as in copper insulated with PVC—but entirely geometrical."[65]

While Iijima has used buckytubes to cast metal wires thinner than DNA, others in Japan and the United States have discovered how to build nanotubes out of various *self-assembling* molecules. For ex-

Figure 1.15 Computer simulation of a double-shell carbon nanotube. (Figure courtesy John Mintmire, Theoretical Chemistry Section, Naval Research Laboratory.)

ample, Akira Harada, a chemist at Osaka University, has assembled tubules 1.5 nm in diameter from cyclodextrin, a glucose derivative.[66]

Reza Ghadiri, a chemist at the Scripps Research Institute in La Jolla, California, has developed a class of nanotubes built from groups of amino acids, or peptides (figure 1.16, plate 3). "For the first time," claims Ghadiri, "tubular structures on the molecular scale can be used in biological settings. . . . These nanotubes may be able to

Figure 1.16 (Plate 3)   A polypeptide organic nanotube in an extended polypeptide lattice. (Courtesy Reza Ghadiri, The Scripps Research Institute.)

form molecular channels, self-assembling inside cell membranes and acting like junctions for transferring molecules into and out of, or between, cells. . . . They're like little test tubes in which we can perform reactions or confine the growth of materials placed inside."[67]

Lipids, which consist of long-chain *aliphatic* hydrocarbons (i.e., the carbon atoms form chains rather than rings), and their derivatives have also been used to construct nanotubes. Because of their aliphatic chains, lipids are insoluble in water but soluble in fat solvents, such as ether, chloroform, and benzene. In an article on lipid tubules, Joel Schnur of the Center for Bio/Molecular Science and Engineering at the Naval Research Laboratory in Washington, DC, argues that self-assembling tubules is an area "ripe for technological breakthroughs." If we can coordinate the expertise of a number of different disciplines—biology, biochemistry, organic chemistry, inorganic chemistry, physics, and materials engineering—Schnur believes that we should soon be able to "design and engineer . . . molecules to form self-assembling structures optimized for specific applications."[68]

Reverse Engineering Biological Motors

Biochemists and molecular biologists have discovered and described thousands of molecular mechanical structures used by living organisms—the most famous being DNA, which holds the information used to construct protein molecules. One family of protein machines under intense scrutiny is the "motor molecules" that move tiny components about within the cell. The interior of all living cells, from fungi to humans, is laced with dense cobwebs of microtubules that provide a network of 25-nanometer-wide train tracks for hauling molecular equipment from place to place. These molecular cargo trains "play a role in many of the cells' most fundamental activities. They may help orchestrate the dance of the chromosomes as they separate into the two daughter cells during cell division. . . . They guide the migration of the small membrane-bound vesicles that carry the enzymes that synthesize neurotransmitters to the nerve terminals where the transmitters are made and released. And they may shuttle into place the protein filaments needed for the assembly of large internal cellular structures such as the endoplasmic reticulum, which is where many proteins are assembled."[69]

Using "optical tweezers"—a device that pins molecular structures in place with beams of light—Steve Block, a biophysicist at the Rowland Institute for Science in Cambridge, Massachusetts, has discovered how kinesin molecules move along microtubule fibers (figure 1.17). Although it remains unclear exactly how kinesin consumes ATP to turn over its engine, Block's research group has "shown that a single kinesin molecule takes an 8-nanometer step. . . . [This] result dovetails nicely with recent research from other labs showing that kinesin molecules naturally sit on microtubules at 8-nanometer intervals, a gap corresponding to the spacing of some of the smaller proteins of which microtubules are formed."[70]

Biomolecules for Computation

Another reverse-engineering effort that is close to producing functional applications is the exploitation of biomolecules as logical units and memory elements in computational devices. Robert R. Birge, Director of the Center for Molecular Electronics at Syracuse University and an editor for the journal *Nanotechnology,* has developed a technique that harnesses the reaction of the light-harvesting molecule *bacteriorhodopsin*—whose femtosecond response to photons is similar to that of rhodopsin—to store and manipulate digital information. Using two lasers that operate at distinct wavelengths, Birge is able to "read" and "write" data by, respectively, interrogating and flipping the configuration of bacteriorhodopsin molecules. Birge has successfully built three-dimensional, light-addressed memory modules that store 18 gigabytes of data in a block measuring 1.6 cm × 1.6 cm × 2 cm. Birge states that "our current storage capacity is well below the maximum theoretical limit of 512 gigabytes for the same 5–cm^3 volume."[71]

As Birge focuses on bacteriorhodopsin, Jonathan Lindsey, a chemist at Carnegie-Mellon University, is investigating another light-

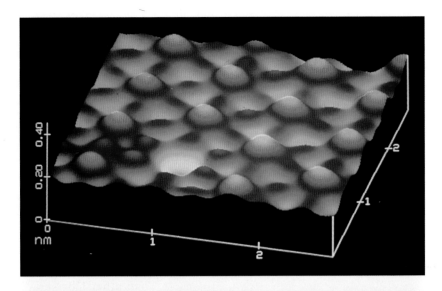

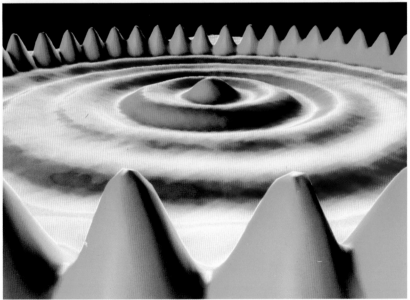

1   Scanning tunneling microscope (STM) image of a single atom defect in an iodine adsorbate lattice on a platinum surface, 2.8 nm square. (Courtesy Digital Instruments, Santa Barbara, California. Photo captured by a NanoScope® scanning tunneling microscope.)

2   Scanning tunneling microscope (STM) image of 48 iron atoms in a circle on a copper crystal surface. (Courtesy IBM Corporation, Research Division, Almaden Research Center.)

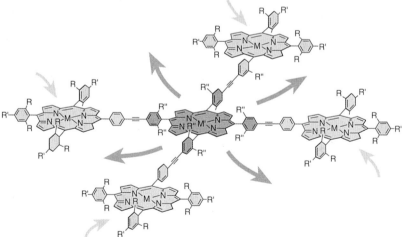

3 Computer graphic simulation of a polypeptide organic nanotube in an extended polypeptide lattice. (Courtesy Reza Ghadiri, The Scripps Research Institute.)

4 A porphyrin pentamer gathers bluish wavelengths of light at its periphery and emits reddish light at its core. (Reprinted with permission from *Science*, 10 September 1993, 1388. Copyright 1993 American Association for the Advancement of Science.)

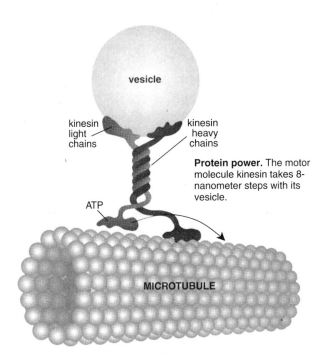

Figure 1.17 Kinesin, a 60-nm molecule, is a biological engine for intracellular transport. (Reprinted with permission from *Science,* 27 August 1993, 1112. Copyright 1993 American Association for the Advancement of Science.)

harvesting molecule: *porphyrin* (figure 1.18, plate 4) Porphyrin is a precursor of both the heme molecule (which binds oxygen in animal blood) and chlorophyll. In plants, chlorophyll arrays are extremely complex. Nobody has been able to figure out their structure or precisely how their configuration relates to their light-capturing function. In an effort to study a simplified version of the chlorophyll mechanism, "Lindsey and his team built porphyrin pentamers, each made of one central porphyrin molecule flanked on each of its four sides by another porphyrin molecule. . . . The outer quartet of porphyrins, they found, absorbs light of specific wavelengths and rapidly transfers the light energy to the core porphyrin, which vents the energy input by fluorescing." Although this research is still exploratory, Lindsey anticipates using porphyrin structures to to build

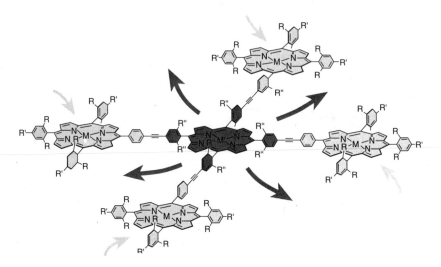

Figure 1.18 (Plate 4) A porphyrin pentamer gathers bluish wavelengths of light at its periphery and emits reddish light at its core. (Reprinted with permission from *Science,* 10 September 1993, 1388. Copyright 1993 American Association for the Advancement of Science.)

"molecular information-processing devices" based on the molecular absorption and emission of photons.[72]

Other molecules are also being pressed into service as computational elements. At Queen's University, Belfast, in Northern Ireland, a team led by Prasanna de Silva reports "the fabrication of a single molecule that behaves as an 'and' gate in a logic circuit. The molecule, an anthracene derivative called benzo-15-crown-ether-aldehyde, fluoresces, or emits light of one wavelength, when exposed to light of another. The molecule can function as an 'and' gate because it reacts differently to two inputs: hydrogen ions and sodium ions. The intensity of this molecule's fluorescence varies, depending on whether a signal comes from the hydrogen channel, the sodium channel, or both. When both channels provide input, the molecule radiates at a stronger intensity, clearly signaling that channel 1 and channel 2 are both on. This technology offers the promise that single molecules could replace whole electronic components, such as transistors. . . . In theory, a cluster of molecules could replace an entire computer chip."[73]

In any attempt to engineer—or reverse engineer—useful molecular structures, a primary hurdle is the ability to predict, from first principles, the atomic machinations and the macroscopic properties of designed, or modified, molecules. Considering the work of Chunming Niu, a chemist at Harvard University, Marvin L. Cohen, a physicist at the University of California, Berkeley, and the Material Sciences Division of the Lawrence Berkeley Laboratory, affirms that we have cleared this hurdle. Niu's research team successfully designed and then created a material that is at least as hard as (has a bulk modulus greater than or equal to) diamond, which had been the hardest known material. Cohen writes that "the confirmation of theory implied by the measurements of Niu *et al.,* indicates that we have entered an era in which it is possible to use theory to design materials with predictable properties."[74]

Directed Evolution of Molecules and Software

In contrast to the "rational" design of molecules, as pursued by Cohen and Niu, some scientists have found ways to use Darwinian selection to *evolve* molecular structures. "Researchers begin with a single molecule, selected for its potential to do some useful chemical task. They make millions or billions of copies of the molecule, each with a slightly varied structure. Then they launch a talent search by making members of this 'population' compete at a task, such as binding to another molecule. They discard those that fail and replicate those that perform well. By repeating this process again and again, each time selecting the best in that generation, scientists . . . evolve a molecule exquisitely adapted to do exactly what they want."[75] For example, Gerard F. Joyce, a chemist at Scripps Research Laboratory in La Jolla, California, and others used directed evolution to produce an RNA enzyme, or *ribozyme,* that would cut strands of DNA, something it does not naturally do. Using a technique developed in the mid-1980s, *polymerase chain reaction* (PCR), Joyce's team generated copies of the original strand of RNA while introducing mutations into each generation. They selected the best performers from each generation and, after twenty-seven generations, their hothouse ribozyme was 100,000 times more effective than the original. Because PCR is able to copy only strands of DNA and RNA nucleotides, this method does not provide a general purpose technique for evolving molecules, but it clearly demonstrates the power of directed evolution.

Darwinian mechanisms are also at the forefront of software design. Stephanie Forrest, a computer scientist at the University of New Mexico, Albuquerque, writes that "genetic algorithms are a search method that can be used for both solving problems and modeling evolutionary systems. With various mapping techniques and an appropriate measure of fitness, a genetic algorithm can be tailored to evolve a solution for many types of problems."[76] Genetic algorithms operate by creating an initial population of binary strings in a computer memory; then the strings are tested for their capacity to execute a given function. Depending on the ability of an individual to perform, it is allowed to "procreate" with other successful individuals and produce offspring that must face similar evaluation. Darwinian functions such as variation, selection, and inheritance have been shown to be useful techniques for evolving binary-string representations of functional programs. Indeed, "biological mechanisms of all kinds are being incorporated into computational systems, including viruses, parasites, and immune systems."[77] John Holland, professor of psychology and computer science and engineering at the University of Michigan, and Maxwell Professor at the Santa Fe Institute, points out that genetic algorithms are particularly useful for modeling many "systems of crucial interest to humankind that have so far defied accurate simulation by computer, [including] economies, ecologies, immune systems, developing embryos, and the brain."[78]

Central to evolution-based simulations as well as directed molecular evolution is the problem of replication and, in particular, self-directed replication. Given the complexity of known biological replicators, researchers have imagined that self-replication was inescapably complex. As a result, early theoretical work in this area did not strive for models that might lead to physical realizations. Von Neumann's initial proposal for a self-replicating machine, for example, embedded a general-purpose computer into a two-dimensional *cellular automata* that required 29-state cells.[79] Recently, however, a team led by James Reggia, a computer scientist at the University of Maryland, has shown that "self-replication is not an inherently complex phenomenon but rather an emergent property arising from local interactions in systems that can be much simpler than is generally believed."[80] Extending the work of Chris Langton, who has shown how to build 86-cell, self-replicating, "Q-shaped sheathed loop" cellular automata,[81] Reggia's team has demonstrated a much smaller, but still Q-shaped, two-dimensional structure that requires only five cells. Reggia notes that, "the existence of these systems raises the question of whether contemporary techniques being developed by organic chemists studying autocatalytic systems [in which structural molecules function as templates for their own replication][82] . . . could be used to realize self-replicating molecular structures patterned after the information processing occurring in unsheathed loops."[83]

Molecular Computation

Another challenge facing molecular engineers is the design of new computational architectures. For example, one of the most difficult problems in contemporary high performance computing is heat-extraction. The computational density implicit in molecular machinery will require novel architectures if the machines are to operate at acceptable temperatures.

One proposal is to make the operations in a computer reversible. It would then be possible to design systems that could "harvest" information from the tidal flow of instructions through computational ecologies. This approach would avoid, at least in theory, the thermally costly move of intentionally destroying stored information.[84] Over twenty years ago, in 1973, a computer scientist at IBM, Charles Bennett, presented a description of a reversible computer: "In the first stage of its computation the logically reversible automaton parallels the corresponding irreversible automaton, except that it saves all intermediate results, thereby avoiding the irreversible operation of erasure. The second stage consists of printing out the desired output. The third stage then reversibly disposes of all the undesired intermediate results by retracing the steps of the first stage in backward order (a process which is only possible because the first stage has been carried out reversibly), thereby restoring the machine (except for the now-written output tape) to its original condition."[85] Reversible computation has not been pursued with microtechnological systems, primarily due to the memory requirements for storing intermediate data, but for molecular computing it may well be essential. Indeed, Ralph Merkle, head of the Computational Nanotechnology Project at Xerox PARC, claims that "reversible logic will dominate in the twenty-first century."[86]

Exploring another novel approach to computation, John Ross, a chemist at Stanford University, and his team have "shown that a network of chemical-filled beakers can perform one of computing's hardest tasks—recognizing a pattern."[87] That is, groups of interconnected beakers, organized into a *neural net,* can perform logical operations by exploiting various chemical equilibria established among the containers.[88]

In related work, Jean-Pierre Banâtre and Daniel Le Métayer, computer scientists at the National Institute for Applied Sciences in Rennes, France, have developed a programming technique based on unordered collections of data objects, such as nested containers of solution-based molecules. This technique is based on the logical notion of a *multiset,* which is the same as a set except that multisets can contain multiple occurrences of the same element. (Multisets are sometimes referred to as *bags.*) Banâtre and Le Métayer suggest that "an intuitive way of describing the meaning of a GAMMA [General Abstract Model for Multiset MAnipulation] program is the metaphor of the chemical reaction: the set can be seen as a chemical solution, function R (called the reaction condition) is a property to be satisfied by reacting elements, and A (the action) describes the product of the reaction. The computation terminates when a stable state is reached, that is to say, when no elements of the set satisfy the reaction condition."[89] While demonstrating that the machinations of multiset transformation provide a powerful technique for describing the operations of parallel-processing systems, Banâtre and Le Métayer emphasize the effectiveness and completeness of their model, rather than any practical realization of it. Indeed, they claim they are "not concerned with implementation issues." In contrast, Jack LeTourneau, Chief Scientist at Prime Arithmetics, Inc., has developed an efficient implementation strategy for this kind of "logical parallelism" based on an arithmetic interpretation of finite multisets.[90]

Artificial Cellularization

The notion of hierarchically nested containers is the conceptual foundation of both object-oriented programming[91] and cellular life. And while multisets provide a *logical* tool for describing cellular interactions, recent advances in polymer chemistry are beginning to provide the *physical* tools for encapsulating microscopic molecular systems. Common polymers, such as nylon and polyethylene as well as DNA, consist of linear chains of simple molecules, or *monomers,* which have at least two reactive sites available for bonding. Samuel Stupp, a chemist at University of Illinois, has now created *two-dimensional* polymers (figure 1.19). In a 21-step process, rodlike molecules with two reactive sites, one at the end and one in the middle, self-assemble into 10-nm thick films. "It is perhaps easiest to understand how these precursors are assembled if one imagines that they are sharpened pencils. The eraser corresponds to the reactive end, and the brand name stamped on the pencil represents the central reactive site. The 'brand name' encourages the pencils to align side by side in the same direction. The pencils therefore form a layer with the erasers on one side and the points on the other."[92]

Such two-dimensional polymer films, which easily extend over several square microns, should be able to provide semipermeable membranes for encapsulated and connected populations of evolving molecular components. Here we might listen to Harold Morowitz,

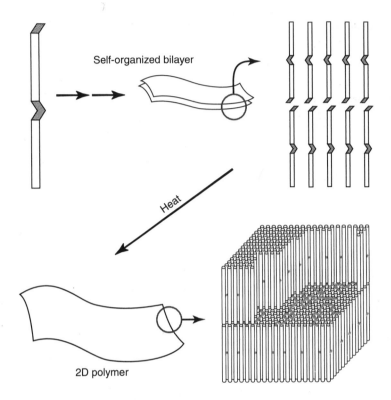

Figure 1.19 Two-dimensional polymer sheets made from rodlike precursors with two reactive sites. (Reprinted with permission from *Science,* 1 January 1993, 43. Copyright 1993 American Association for the Advancement of Science.)

professor of biology at George Mason University, who has argued that "it is the closure of an amphiphilic bilayer membrane into a vesicle that represents a discrete transition from nonlife to life. . . . It seems likely that . . . the emergence of deoxyribonucleic acids, transcription, and other elaborations came later. . . . The first radiation leads from the earliest vesicles to the universal ancestors. At this stage biogenesis is over, and the origin of species begins. The rest is history."[93]

History is also what this book has become. That is to say, it is out of date. As you read this, our molecular knowledge is alive and growing. Books can never depict this cutting edge. If you want to know what's happening *now,* get on the Internet.

To connect to the Internet, you need a computer and a modem. If these resources are not readily available, contact your local town hall, library, or school for information. Once on-line, any of the following World Wide Web (WWW) sites can serve as an entry point to the current state of the art of nanotechnology. Web search engines, such as Alta Vista (*http://www.altavista.digital.com/*) or Lycos (*http://www.lycos.com/*), are also useful if you are interested in a particular area of nanotechnological research; searching on "nanotechnology" alone will turn up thousands of web pages.

Academic sites include the National Nanofabrication Users' Network at Stanford University (*http://snf.stanford.edu/NNUN/*), the University of Southern California's Laboratory for Molecular Robotics (*http://alicudi.usc.edu/lmr/molecular_robotics_lab.html*), the Materials and Process Simulation Center at the California Institute of Technology (*http://www.wag.caltech.edu/*), the Nanostructures Laboratory at MIT (*http://www-mtl.mit.edu/MTL/NSL.html*), the Molecular and Electronic Nanostructures group at the Beckman Institute for Advanced Science and Technology at the University of Illinois (*http://www.beckman.uiuc.edu/themes/MENS.html*), the National Nanofabrication Facility at Cornell University (*http://www.nnf.cornell.edu/*), the Nanomanipulator Project in the Department of Computer Science at the University of North Carolina (*http://www.cs.unc.edu/nano/etc/www/*), the Center for Molecular Design at the Institute for Biomedical Computing at Washington University in St. Louis (*http://wucmd.wustl.edu*), Ned Seeman's laboratory at New York University (*http://seemanlab4.chem.nyu.edu/homepage.html*), and Tom Schneider's Laboratory of Mathematical Biology at the National Cancer Institute (*http://www-lmmb.ncifcrf.gov/~toms/toms.html*).

Other sites include the IBM Almaden Research Center Visualization lab (*http://www.almaden.ibm.com/vis/vis_lab.html*), the Computational Molecular Nanotechnology laboratory at NASA Ames Research Center (*http://www.nas.nasa.gov/NAS/Projects/nanotech-*

nology/), The Foresight Institute (*http://www.foresight.org/*), Ralph Merkle's Computational Nanotechnology laboratory at Xerox PARC (*http://nano.xerox.com/nano/*), The Molecular Manufacturing Shortcut Group, which promotes "the development of nanotechnology as a means to facilitate the settlement of space" (*http://www.music.qub.ac.uk:80/~amon/IslandOne/MMSG/*), Molecular Manufacturing Enterprises Incorporated, a seed capital investment firm (*http://www.mmei.com/*), Nanothinc, Inc., an information company seeking to trade at the development horizon of nanotechnology (*http://www.nanothinc.com/*), the "sci.nanotech" Internet newsgroup archives, which are maintained by Josh Hall, the creator of utility fog (*http://athos.rutgers.edu:80/nanotech/*), and the *Nano-Technology Magazine* home page (*http://planet-hawaii.com/nanozine/*).

For three-dimensional molecular images, see CambridgeSoft's Chemfinder Database (*http://chemfinder.camsoft.com/*), for a complete chemical reference, see the Cambridge Structural Database (*http://csdvx2.ccdc.cam.ac.uk/csd.html*), and for a detailed index of chemistry sites on the Web, see Mark Winter's ChemDex at Sheffield University (*http://www.shef.ac.uk/~chem/chemdex/*). For an index of artificial life publications available on the Web, see Ezequiel Di-Paolo's bibliography at the University of Sussex's School of Cognitive and Computing Sciences (*http://www.cogs.susx.ac.uk/users/ezequiel/alife-page/alife.html/*).

I Mostly Inside

■ 2 In-Vivo Nanoscope and the "Two-Week Revolution"
Ted Kaehler

Neither the naked hand nor the understanding left to itself can effect much. It is by instruments and helps that the work is done, which are as much wanted for the understanding as for the hand.
—Francis Bacon

The reason we are on a higher imaginative level is not because we have finer imagination, but because we have better instruments.
—Alfred North Whitehead

☐ In-Vivo Nanoscope

In a cage in front of us is a pregnant mouse. To one side, two women are arguing about the meaning of a live video picture that shows molecules interacting in a cell inside the mouse. The pregnant mouse is cleaning her fur and occasionally scratching at a white thread that emerges from the side of her abdomen. The thread enters the skin surprisingly cleanly, because its surface exhibits mouse skin-cell markers and is accepted as normal by the surrounding tissue. The thread travels under the skin and then dives into the muscle of her abdomen, again accepted as normal tissue by the mouse's cells. Though the thread is small in diameter, it has a complex structure of fiber optic links and power cables.

The thread eventually crosses into the womb of the pregnant mouse and into the circulatory system of a mouse embryo. In the thymus gland of the embryo, the thread is firmly anchored to the wall of a blood vessel and begins a succession of narrowings until it is no wider than a blood cell. The tip of the thread is just withdrawing from an immune system T cell and poised to enter another cell.

The tip left no wound when it withdrew, for its surface carried the same phospholipid pattern as the membrane of the cell. As it enters the next cell, it becomes an integral part of that cell's membrane. As before, the thread is accepted by the membrane in the same way that the membrane accepts large proteins that also cross it in their role as molecular pumps.

How is this thread accepted inside the body of an animal without rejection? The sides of the thread consist of a series of flaps that can lift to expose a variety of different manufactured surfaces. The tip of the thread is an octopus-like array of sensors on stalks. Each is no thicker than a large protein molecule and waves constantly, probing its watery surroundings. At the base of each stalk, motors whir to generate movements, and simple mechanical "electronics" collect the data from the sensors. The sensors "feel" their environment using force, electric charge, moving magnetic fields, and a variety of hydrophobic and hydrophilic probes. The sensors are small-scale versions of the rich family of scanning-probe microscopes. The data are sent back up the fiber optic channels in the thread for processing by computers in the lab.

The thread and its accompanying computer and displays form the in-vivo nanoscope, a vital tool for the hands-on investigation of biological processes inside the cells of a living animal.

In the room with the mouse cage, the scientists argue about the process by which the T cells are being "educated" in the thymus of the mouse embryo. On a dozen computer screens around the room, live video pictures display false-color views of what the tip of the thread is sensing inside the mouse cell. Just as Landsat[1] images are often displayed, each monitor combines the data from several sensors into a single color picture. The colors are not "true" because the sensors do not see light. Instead, a computer colors the pictures "falsely" in order to highlight interesting details.

The two women are doing research on the immune system. It has been known for quite some time that immune system cells have to be educated about what is the body's own tissue and what is foreign matter that should be destroyed. But the exact mechanism for how this happens in the thymus gland is not well understood. If the details of the ways in which the body's T cells distinguish between friend and foe were known, the secrets of arthritis, diabetes, and other auto-immune diseases might be unlocked.[2]

Before the development of the in-vivo nanoscope, research on the chemical processes inside living cells was extremely difficult. What scientists learned in the days before the nanoscope is amazing, considering how indirectly the knowledge was obtained. To study a process inside a live cell, one often had to fuse that type of cell with cancer cells so they would grow without restriction in the lab. Then, while keeping finicky cell cultures alive, one had to introduce dyes or radioactive chemicals that would bind to the portion of the cell under scrutiny. Getting all these indirect methods to work required a good idea of what one was looking for. It was like drilling a hole in a bank vault and using a piece of wire to poke around in order to find out how the lock worked, or deducing the layout of the complex gears inside without being able to see them.

In the lab with the mouse, the two scientists continue to argue, "Well, if that's true, then we should see pieces of DNA actually being spliced here, but we're not."

"You just aren't looking in the right place. Look over there." The first scientist looks at several false-color displays and concentrates on moving the controls for a minute. "There—look at that."

"Yes, but how do you know that that's really what we're seeing? Here, let's put some better filters on that image . . ."

With the direct visualization made possible by the in-vivo nanoscope, biologists can learn in an afternoon what formerly took years of groping with frustratingly indirect methods. A new generation of drugs and therapies is emerging as a result of this leap in understanding of the processes within normal, living cells.

☐ **Pre-Assembler Nanotechnology**

The in-vivo nanoscope is the product of an awkward stage in the development of nanotechnology. It required a building full of equipment and graduate students to construct the nanomachines at the tip of the nanoscope. Most of the "thread" and its surface are made with conventional technology. The computers with the intelligence for processing the images are conventional desktop machines sitting in the lab at the other end of the thread. True assemblers have not yet been made.

Assemblers, universal nanomachines, are nanoscale machines that can place atoms with precision and bond them in any possible pattern by following a stored design, or program. At the time of the invention of the in-vivo nanoscope, a building full of equipment can make a nanomachine, but no existing nanomachine will immediately be capable of making another nanomachine by itself. A true assembler must contain a computer, it must be able to "see" and make sense of its surroundings, it must be able to propel itself and navigate, and it must be able to handle and bond atoms of various kinds. Assemblers will be built someday, but not in the same decade as the simple nanomachines located at the tip of the in-vivo nanoscope.

☐ **The "Two-Week Revolution"**

Never in history has a technology been so clear in our vision yet so frustratingly out of reach as is nanotechnology in the 1990s. We know a lot about what can be done with nanomachines, but we cannot build them now. If we had a single assembler, we could order it to build another one. Those two could produce more, until any desired number were produced. Then we could tell them to build other things. A great flourishing of new nanotech machines would follow—as fast as we could design them.

But there is a technological bottleneck before we can build the first assembler. We have been accumulating plans for pieces of nanomachines, starting with the first bearing designs presented in 1987, but we remain unable to build them.[3] It may be decades before we can actually build the first working assembler. In that time, we will accumulate many levels of designs, leading to a spectrum of useful nanomachines. When the first assembler works, a stampede of working machines could follow. All we have to do is dust off the designs and tell the newly created assemblers to go to work.

This sudden surge of working nanomachines has been called the "two-week revolution."[4] In the first two weeks after the assembler breakthrough, the world may change radically. For some, this is not a metaphor but a prediction of great change in a matter of days. Entire new systems of fully functional technology will emerge, ready to transform the world.

☐ **What, no learning curve?**

The two-week revolution will not happen that way. Development will occur gradually after the first assembler works. Using the development of computers as a parallel, we can see that design and construction must go hand in hand. If we accumulate a backlog of designs, they simply will not work at first. Steady and magnificent technological progress has been the hallmark of the last half century. It is a testament to the power of the product development cycle: design something, test it, learn what needs to be changed, modify the design, and repeat the cycle. The learning curve for computer chips shows that the price of a given chip falls in relation to the total number of units shipped since the product first came out. The ability of an industry to learn how to make a product depends very strongly on how many units have been made. Hundreds of little corrections and small tricks help lower the price of a unit in the next production run. Sheer experience is the most important factor. Many products from VCRs to photocopiers exhibit the same kind of learning curve and attest to the power of feedback and gradually modifying the design.

Consider the designers of a nanomachine. They design the device, using software on personal workstations. As soon as they think they are done with the design, they choose "Build" from a pop-up menu, and a group of assemblers in a tiny capsule fabricate the new device. Suppose that the building process is absolutely perfect. The nanomachine is assembled exactly the way the blueprints say it should be. Will it work the first time?

We have a wonderful example of a perfect medium to learn from: computer software. When the software designers are finished writing a program, they ask the system to compile, and a software "machine" is built—and built perfectly. The only thing that can go wrong are "bugs." A bug occurs when a program does exactly what the programmer wrote down, but this is not the intended result. One might think that this would not be much of a problem, but it is. Software must be tested over and over in order to discover all of the subtle bugs. The length of time needed to test and debug a large software system is notoriously large.

☐ **Bugs, Bugs, Bugs**

Why does software, the most perfectly manipulable of all media, still take such a long time to debug? It is because software is very dense with ideas. A small amount of software code can express a complex idea. The programmer weaves many such pieces into the complex fabric of a large system. There are lots of subtle interactions among the parts of the system. Also, there is a great contrast between the static, written form of the program and the active movement of data in the running system. Programmers have to imagine what will happen dynamically when the computer executes the program. Often they picture it incorrectly. They do not see the consequences of the many threads interacting in different combinations.

As with any complex system, the same bugs and misunderstandings face the designers of a nanomachine. Assuming the design is executed accurately, untested nanomachines will have the same kind of bugs as untested software. (Not counting the fact that a nanodevice will contain a large amount of software running on its own on-board computer.) As with software development, the major problems will be caused by misunderstandings on the part of the designer. One misreads the specification of a module that someone else designed and tries to use it in the wrong way. One forgets an assumption made a month before—that a certain module would never be used in a certain way. One pushes a module beyond its design limits by overloading it in some way—making it drive too many other things, making it go too fast, not holding input signals steady for a long enough time, and so on. One uses someone else's module correctly, but in a way that exposes a bug that no one has encountered before. Anyone who has debugged a computer program knows these frustrations intimately.

In addition to software bugs, there is a broad class of "process errors." In the first two weeks after the first assembler is built, the art of building nanostructures will not yet be flawless. The assembler may make a mistake and install the wrong atom at some location. It may not be able to reach the place where the next atom needs to go. It may be unable to make an element bond properly. The structure it is building may become unstable when only half finished, and react chemically or fold up in a strange way. All of these errors can be corrected; assemblers will eventually work wonderfully well— just not in the first two weeks. Many cycles of design and test will have to be performed by humans to get everything working.

As one thinks about this monumental "design-ahead" project and tries to imagine planning for every conceivable eventuality, one comes to appreciate the power of fast turnaround. The technique of finishing a design, testing it, figuring out what went wrong, and fixing the design is a very powerful tool. One of the main reasons that Paul MacCready and his crew were able to build the first human-powered airplane is that they had quick turnaround. Before the Gossamer Condor, many others had tried, but they all spent months painstakingly building a beautiful plane and then crashing it in a few seconds. MacCready designed his plane to be fixed. His plane also crashed almost every time it flew, but it could be repaired or rebuilt easily. Sometimes it flew again just twenty-four hours after a crash. This short design cycle enabled MacCready's crew to make rapid progress on improving the design.[5]

This is also true of commercial products. Advances in camcorders, Walkmans, and personal computers have been much greater than the advances in space flight over the last ten years. This is largely because the design-cycle time for a consumer item is months, and the design cycle for spacecrafts is closer to a decade. Yet we hear the proponents of the "first two weeks" saying we will be able to do it with infinitely slow turnaround, without *any* testing cycles before the first effective assembler is built.

☐ A Counterargument

Those who believe in the two-week revolution argue that computer simulations can be detailed and accurate enough to find and correct virtually all the bugs before any working machine is built. They note that similar problems are faced by computer-chip designers. It can take up to six weeks to run a batch of wafers through a chip-making factory. Because of this time lag, it is very important to make sure a newly designed chip works before it is sent off to be built.

Chip designers use simulators to find out what the chip will do. Working with computer files that describe the layout of the chip, they submit them to programs that extract the circuit from the design. Simulators try out the circuit, and the designers compare the results to what they think the chip should do. One kind of simulator tests to see if individual transistors will work as expected. Another looks at the timing of signals traveling long distances across the chip. Another assumes that everything works on that level and tests the logic of the 1's and 0's. Another assumes that small sections of the chip work and tests if the chip as a whole produces the right answer. Such modular testing techniques might allow us to simulate all important aspects of nanomachines before they are built. Thus, when we are finally able to build them, we will have confidence that they will work as planned.

Indeed, there is a famous example of this approach working. During World War II, the Manhattan Project built a gas-diffusion separation plant in Oak Ridge, Tennessee. Its job was to separate the light isotope of uranium from the heavy one, which no plant had ever been built to do before. The principle upon which it ran was new. The gas was corrosive, and the plant had to resist it on every surface that touched the gas. In addition, the gas and the plant around it were so radioactive, repairs or rebuilding would be extremely difficult. There was no time to build a small-scale pilot plant.

Stephane Groueff describes the challenge. "Barrier problems, pump problems, corrosion problems, cascade problems. One of the major difficulties was the extraordinarily corrosive power of uranium gas; it ruled out the use of practically every known metal. Another one was the requirement of airtightness. Such a requirement exceeded—by far—anything that had ever been conceived before, even in small systems. To incorporate such airtightness on a gigantic scale, in piping hundreds of miles long and through two thousand cascade stages, seemed an impossible dream. The gaseous-diffusion system could be ruined by even an infinitesimal leakage. Such minute leakage, however, seemed unpreventable. . . . No existing valve, no welding technique and no pipe joint could guarantee such optimum airtightness."[6] The plant had to work the first time. It did.

Some chip designers have had the wonderful experience of having a complex chip work the first time. Many others have had to test the chip, think hard about what went wrong, make changes, and submit the design for fabrication again. And again.

Projects to build large artifacts suffer from the same it-has-to-work-the-first-time problem. When you cannot build prototypes and properly test them, you have to go to extraordinary lengths to make the machine work. The space shuttle flew with humans on board the first time. However, the amount of design effort, and the huge number of design change orders, attest to the difficulty of working this way. And indeed, the design of the shuttle is flawed in many ways.

☐ **Tools for Design-Ahead**

We know some very powerful tools for doing good design without testing it. "Design rules" are an especially powerful technique. A transistor in the middle of a computer chip can be varied in many ways. Different shapes of transistor, for example, will have different speed and power. But chip designers often give up that flexibility and agree to follow a set of rigid design rules. The rules only allow certain kinds of transistors, sometimes as few as three simple configurations. These transistors are slower than optimum and have a lot of restrictions on how far they can send their signals, and so on.

Why would designers give up all that flexibility and produce a design that is worse than it could be? Because they know it will work. If you were to design your own transistors, you would make mistakes. It is a complicated and specialized business. The set of official design rules are certified by the people who manage chip fabrication. When designers follow these rules, they are more or less guaranteed that the transistors will work.[7] A significant development in nanotechnology will be a good set of design rules. Such rules are rarely the best, but they can be used with assurance.

Another powerful tool is the idea of modules. A module is a "black box" of machinery whose inputs and outputs are carefully described. A designer can use someone else's module in his or her design without understanding how it works. The designer needs only to know how to hook it up and what it does. Cooking offers several examples. If you are preparing a recipe that calls for mayonnaise, it's very convenient to buy mayonnaise from the store, rather than making it yourself. You could, of course, prepare it from scratch, but you would need to know quite a bit about mayonnaise. It's often more practical to put that thought and effort into preparing the dish at hand. Buying the "mayonnaise module" off the shelf allows you to focus on the final culinary design.

Dividing a machine into modules greatly simplifies the design process. The "owner" of one module can become an expert in making it work better and can debug it thoroughly. Many people can use a well-designed module, and the number of things they need to think about is greatly reduced. The result is a more reliable overall design.

☐ **Questions of Degree**

Simulation is a powerful tool, but it has its limits. Yes, it is theoretically possible to have designed an entire set of personal computer software in 1950—before the first commercial computer was shipped. It could have been all worked out on paper and in people's heads, but the number of levels of new ideas required is staggering. Today's software tools did not appear fully formed at the dawn of the computer age; their evolutionary emergence occurred over time. The advantage of having a spreadsheet on a personal computer was not obvious even after it was invented. At least one major company was offered the rights to VisiCalc™, the first spreadsheet program for a personal computer, and refused. Nanotechnology's antici-pated two-week revolution is akin to predesigning transistors, chips, memory systems, operating systems, computer science, several generations of programming languages, personal computers, and ap-plications in the year 1950. We are asked to discover and appreciate the amazing utility of a spreadsheet, when no person has ever expe-rienced an hour in front of a personal computer.

Another issue is the cost of massive design-ahead. Each layer of design that remains simulated but untested adds to the uncertainty of the device actually working. Investors will have a limit to the number of untested assumptions that can be believably piled on top of each other and may wait until the first assembler proves itself in the laboratory before they invest massive amounts of capital.

Patents are another consideration that will influence how much design-ahead is done. No one knows the extent to which designs for nanomachines will be patentable, and it may be hard to patent something that is impossible to build today. In general, patents on molecules are not given unless the molecule has been synthesized, however, the molecular parts in a nanomachine are much more like mechanical parts, and a very good computer simulation is arguably a working model of the actual machine. If patents are granted before it is possible to build the device, venture capitalists will be much more willing to pour money into design-ahead.

While some nanodevices will be designed before the first assem-bler is working, they will not be whole products ready to do things in the real world. They will not be the things described in the rest of this book. Those will come later. The sheer number of new layers of design in mature nanomachines make it very unlikely that all of it will work the first time. The practical details of getting a design to actually work in the real world is a human endeavor. Humans need time to understand ideas, communicate, get used to new things, and get the right social organizations in place. Experience is the best teacher. While simulation can go a long way, we can be sure that the real thing will contain plenty of surprises.

☐ Conclusions

The two-week revolution will not happen. Two weeks after the first assembler works, it will be in the shop for repairs. And not many of the things that it built in those two weeks will work either. The pervasive use of assemblers in our lives depends on the development of several new fields of study and entire new layers of infrastructures. It will be a human endeavor operating at human speeds. It won't happen without thousands of cycles of experimental feedback, and certainly not in the first two weeks.

☐ Acknowledgment

The author thanks the attendees of the "Assembler Multitude" meetings of the Nanotechnology Special Interest Group of Computer Professionals for Social Responsibility (CPSR), Palo Alto, California, for contributing and discussing many of the ideas presented in this chapter.

■ 3 Cosmetic Nanosurgery

Richard Crawford

Beauty is truth, truth beauty—that is all ye know on earth, and all ye need to know.
—John Keats

☐ The Beauty Business

One of the earliest and most rudimentary applications for nanotechnology may well be in what many might consider a frivolous cause—the alteration and enhancement of human appearance, otherwise known as the beauty business. Why? Because the technology can be relatively simple, there is a great demonstrated demand, and vast sums of money can be made. Even with the crude techniques available today, the cosmetics industry is thriving and lucrative. Once the means are at hand to actually perform safely, painlessly, and inexpensively the miracles that advertisements now promise, the race will be on.

☐ The Market Potential

How big is the beauty business? That depends on how you define it, but a few representative figures will provide a rough idea. According to various sources, the worldwide gross volume for toiletries in 1990 was in the range of $14 to $18 billion. That's just for traditional personal hygine products such as powders, sprays, perfumes, and deodorants.

The appearance-enhancement market goes far beyond that. The diet industry is said to gross $33 billion annually. A recent survey by *Glamour* magazine found representative young American women spending from $550 to $7400 a year on their looks, including such items as contact lenses, gym sessions, and bicycles for exercise. And, according to a recent *Wall Street Journal* article, the total spent for breast implants, hair transplants, facelifts, tummy tucks, and other surgical repair operations comes to about one billion dollars a year, most of it for women but a growing percentage (about 30 percent by some estimates) for men.

These expenditures are for goods and services that, by and large, perform far less miraculously than their advertisements would lead one to believe. Indeed, beauty-related goods and services are named in ways that suggest a level of performance unattainable in practice, such as "permanent" waves that last a few months at most.

Far more pervasive, and insidious, are ads that use gorgeous models to imply—without really saying—that using the product will make you look beautiful too. Surely no one consciously believes such advertising, but the implication is there. The subconscious naively accepts and reacts.

No one knows what people would pay for products that really did transform one's appearance. It's reasonable to suppose, however, that it would at least equal and perhaps far surpass what they now pay for cover-ups and nostrums.

And this is not the full extent of the market for cosmetic nanosurgery. Once this technology begins to mature, it may become possible to address beauty needs that previous generations simply bore without hope or recourse. Typical examples might include completely scarless facial surgery, rebuilding poorly set bones to normal contours, and repairing nerve damage that affects facial muscles.

☐ **Applications of Cosmetic Nanosurgery**

Hair Color

To get an idea of cosmetic nanosurgery's potential, consider a hair-color product as a simple example. This might take the form of a rudimentary, nonreplicating nanomachine—little more than a carefully designed molecule—that simply circulates in the bloodstream until it finds itself in a hair follicle in the scalp. Then it inserts itself into one of the follicle's melanocytes, the cells deep within the hair-root bulb that color the hair, where it benignly but persistently regulates the production or delivery of the pigment melanin. (We'll return later to the question of how to produce such a device and what its properties would need to include.)

Hair color, except in the case of gray hair, depends less on the amount of melanin than on its distribution. Red hair has regularly spaced melanin granules, whereas in brown hair the melanin granules are irregularly spaced. Black hair contains large melanin granules, and in blond hair they are relatively small.

While the effect of a single melanocyte changing to a different mode of melanin production would be detectable only with a microscope, a small syringe could easily contain more such nanomachines than there are melanocytes on your whole scalp. As a result, a brunette could bleach her hair once and never again have to worry about dark roots growing out. She would, in effect, become a blond from the inside out, requiring only an occasional booster shot to replace any nanomachines that get trapped accidentally in the hair shaft and dragged out of the follicle.

Further, just as in normally blond hair, no two strands of hair would have exactly the same shade. (That's how you can tell a bleach job from the real thing.) The color of each hair shaft would depend on how completely the melanocytes had been converted to the new mode, which would naturally vary somewhat from follicle to follicle. And, since hair waviness is also controlled by what happens in the follicle, it might be subject to similar modification with a suitable nanomachine.

Such a hair-color nanomachine could be rudimentary. It would require no on-board computer, just the ability to recognize first a scalp follicle and then a melanocyte. It could do both by binding temporarily to an appropriate surface protein on the target cell. Then it would simply bind to some structure within the melanocyte in such a way as to influence melanin formation. There would be no need for the self-replication machinery an assembler would require;

in fact, self-replication would be a real disadvantage in this and many other applications.

Some sort of self-replicating machinery would be necessary to produce these rudimentary nanomachines in useful numbers, of course. This machinery need not be completely human constructed, however. It should be possible to engineer a number of bacterial strains to produce the various components of these simple nanomachines. These components would then self-assemble when they were mixed together, as virus components do.

A similar nanomachine could be programmed to have the opposite effect—to restart melanin production for those who wish to restore the hair color of their youth. It would probably need to be more sophisticated than the nanomachine we just described. There are several causes of gray hair, so it would need to be "smart" enough to recognize and evaluate symptoms and prescribe a cure. Alternatively, one might design several different varieties of nanomachine, one for each known cause of gray hair, and inject the proper one to treat a particular patient based on the results of testing.

Suppose someone who has taken this treatment changes his or her mind. Suppose gentlemen begin to prefer brunettes, or fashion suddenly decrees that redheads deserve to have more fun. Would the new blonds be stuck with their choice?

Not at all. It should be a simple matter to design a similar hair-follicle-seeking nanomachine to simply turn off the first one. (It would take a different and more sophisticated nanomachine to turn a natural blonde into a brunette.) In fact, both nanomachines could then allow themselves to be incorporated into the growing hair shaft and thus eliminate themselves from the body. This flexibility would also be desirable from a commercial standpoint, giving changing fashions a chance to provide a continuing market.[1]

It has been suggested that gene therapy, already in clinical trials for certain rare diseases, could probably address these same needs and be equally reversible. Gene therapy is also likely to be available long before functional nanomachines appear. However, gene therapy will have major political obstacles and formidable buyer resistance to overcome before it becomes accepted for taking care of anything other than life-threatening diseases.

Skin Color

A similar set of nanomachines, perhaps even simpler because they would have no need to seek out hair follicles, would allow people to control their skin color. These machines could be made to act locally—to deal with birthmarks, liver spots, and freckles for example—or allowed to simply enter whatever melanocyte they encounter anywhere on the body. People could get whatever skin shade—or eye color—they choose, lighter or darker. The social consequences of such innovations are perhaps best left to the imagination. There can be little doubt, however, as to its market potential.

Baldness

Before leaving the subject of hair follicles, it seems evident that there would be a stupendous market for a nanomachine that would restart hair growth in follicles that had turned off. There is probably no branch of the cosmetic industry that has garnered more business with less benefit to the customer than the potions that promise hair restoration. A product that actually worked would be an instant sensation.

Lest anyone argue that hair restoration is an impossible dream, it might be well to mention that not all hair loss is permanent, even now. Chemotherapy and various medical conditions can cause complete hair loss, which reverses itself when the therapy is discontinued or the condition is corrected. And it seems intuitively reasonable that a nanomachine working from inside the hair follicle, right where the action is, should have a better chance of success than any amount of medication trying to fight its way in from outside the scalp.

One problem in developing any such product would be to guarantee that it would function only where one wants it to. No one wants hair sprouting from their forehead or eyelids, for example. (Yes, there are hair follicles on your eyelids. They just produce very short and nearly invisible hairs.) Similarly, women with thinning hair (it happens) wouldn't be interested in a hair restorer that also grows beards. The product would have to include strong safeguards against indiscriminate action.

To accomplish this, nanomachines would need to have some way of "knowing" where they are in the body. However, this need not be nearly as complicated as might be imagined. One simple way might be with subcutaneous injections of a colorless chemical marker—an invisible tattoo—that would gradually seep into the surrounding tissues. This would set up a localized concentration gradient that the nanomachine could recognize as its start-up signal. Once the nanomachine had become attached to a particular follicle it would stay there, indifferent to the gradual dissipation and eventual disappearance of the chemical marker.

Unwanted Hair

Unwanted hair is an ongoing problem for many men and women: hardly anyone really enjoys shaving. A nanomachine that would enter hair follicle cells and stop unwanted hair production once and for all would be welcomed with enthusiasm.

An alternative and possibly simpler way to avoid the need to shave, although it might involve much more mature technology, would be with a one-time external application of a nanomachine-laden depilatory cream. In this case, the machines would attach themselves permanently to the stubble ends and busily convert the protein molecules of each hair stub (almost all keratin) into harmless, odorless gases such as methane, nitrogen, water vapor, and carbon dioxide, which would unobtrusively dissipate. It should be possible to get them to digest each hair stub faster than it grows out, eliminating shaving forever.

Sulfur, which constitutes four percent to six percent of hair by weight, isn't really a troublemaker, even though all of its volatile compounds are smelly (mercaptans, sulfur dioxide, hydrogen sulfide) and some are skin irritants or toxins. Elemental sulfur, on the other hand, is relatively inert. It should be safe to let it accumulate in little crystals until the next washing removed it. The only potentially troublesome nonvolatile constituent of hair protein appears to be phosphorus, which is present in very minor amounts. Phosphorus itself can ignite on contact with air, producing skin-irritating oxides and acids. However, it is unnecessary to carry the breakdown of the hair protein that far. It should be simple to program the nanomachines to leave the phosphorus in a harmless water-soluble organic compound that would wash away in the next rinse.

As a further refinement, these nanomachines could be made to depend on a continuous supply of atmospheric oxygen. For example, the energy to do their work could come from the flameless oxidation-reduction reactions the nanomachines would be promoting. This might provide a way to regulate their activity; whenever the supply of oxygen diminished, the nanomachines would slow down and wait for it to pick up again. This would be likely to happen whenever the end of the hair stub sank far below the skin surface and became covered with sebum (the natural hair lubricant secreted into each follicle by its sebaceous gland). In any case, some such regulator might be advisable to assure that no overenthusiastic nanomachines could work their way deep into a follicle and somehow damage it.

For this scheme to work, it would be important to give the nanomachines the ability to distinguish between hair keratin and the skin keratin, and keep them from harming the latter. One way to do this might be to give each nanomachine at least five other arms (or attachment sites) in addition to the one for manipulating keratin structure, and require them to link up in a way that would be easy on hair and hard on skin. For instance, we might make the manipulator arm inoperative unless the other arms were linked to two adjacent cystine molecules—the predominant amino acid component of keratin—perhaps by recognizing cystine's characteristic disulfide group, and also to at least three other nanomachines. Hair is almost all keratin, whereas skin keratin is mixed in with other components, making it hard for the nanomachines to fulfill the attachment criteria on skin.

Nanomachines that failed to attach themselves in this way would be neutralized with a follow-up rinse. Those already attached to keratin, either on the skin or on a hair stub, would thereby become immune to the rinse.

Any nanomachines that did attach themselves to skin keratin would find slow going in comparison with the others on the hair stubs. They would keep running into nonkeratin components that they would be unequipped to disassemble. Long before they had made much headway, the skin particle they were on would flake off. Once such a nanomachine ran out of keratin to dismantle, it would lose its immunity to moisture and would in all probability come apart before it could attach to another source of keratin.

At the same time, this multiple-attachment requirement would make it possible for each correctly attached nanomachine to stay stuck to the hair shaft even as it cuts away the ground under its own "feet." Whenever one of its attachment sites to the hair shaft became undermined, it would just grope around with the free arm until it found a fresh handhold and then keep digging on the other side.

One obvious precaution would be to prevent this nanomachine from attacking the keratin in fingernails, which appears to be chemically indistinguishable from that in hair. The easy way to do this would be to wear gloves, but that still leaves open the possibility of damage through carelessness. It would undoubtedly be safer to provide a positive deterrent.

The solution might be to make the nanomachines susceptible to attack by some simple solvent such as acetone, a common nail-polish-remover ingredient and something not normally found on the face. This would also make it possible to readjust the boundary of the area within which the nanomachines can act if one wants longer sideburns or decides to grow a mustache.

A Permanent Breath Freshener

A similar topical application would address another personal hygiene concern: bad breath originating in the mouth. As most of us are aware, halitosis frequently results from decaying food particles that perfunctory brushing has failed to dislodge. A useful preparation would bond itself firmly to the tooth enamel and prevent anything else from sticking. It would also actively remove anything that became mechanically wedged.

In this case one would start with a thorough professional cleaning to remove any accumulated plaque and tartar. The dental technician would then paint on a solution teeming with nanomachines that cling tenaciously to every bit of exposed tooth enamel and link up to form a complete, invisible surface film around each tooth. Each of these nanomachines would then go to work disassembling any foreign matter that came in contact with the tooth into harmless and bacteriostatic gases or liquids, which would quickly be removed by saliva.

The same sort of idea could have many other applications such as always-clean windows, self-cleansing dishes, the permanently spotless and odor-free bathroom, and a true antifouling paint for boats. The latter would enjoy a multimillion-dollar market all by itself. For these applications the nanomachines could have much simpler attachment and attack criteria.

To return to the fresh-breath nanomachine, such a system would obviously need to be "smart" enough to recognize gum and tongue tissue and leave them alone. It might also need to be periodically reapplied, as normal chewing and brushing would tend to remove it from exposed surfaces. However, the protection would still be there on the hard-to-reach surfaces, and that's where it's most needed in any case. As an added bonus, the nanomachines could be programmed to construct oil of peppermint (or whatever aromatic essence might be desired) out of the organic molecules they have been dismantling.

Although such a rudimentary nanomachine system would certainly improve mouth cleanliness and go a long way to prevent cavities and tartar buildup, it would not by itself be enough to prevent periodontal (gum) disease if it were used as an excuse to stop gum massaging and flossing. That advance will probably have to await the development of fully capable cell-repair nanomachines, which would conquer all diseases including gum disease.[2] Until then we will have to keep supplying manually the exercise our gums need to keep them vigorous and healthy, a discipline that is necessary because our usual bland diets of refined foods fail to provide the stimulation. One hopes that the promoters of a breath-freshener system will market it as a between-brushing safeguard instead of a dental hygiene cure-all.[3]

Wrinkle Repair

As cosmetic nanosurgery technology matures, it should become possible to undertake more ambitious transformations such as skin rejuvenation. The saying, "Beauty is only skin deep," simply serves to emphasize how very important the skin is.

The skin is a very complex and little appreciated organ, exposed to a wide range of environmental assaults that would simply destroy any of our other vital organs. It resists extremes of heat and cold, dryness and abrasion, biologic, organic, and inorganic poisons, invasive organisms, the force of gravity, and continual bending and flexing. No wonder it begins to look somewhat worn after a few decades!

Total and spontaneous skin renewal will probably have to await the arrival of fully capable cell-repair nanomachines. These would be able to analyze whatever ails a particular cell or tissue component and restore it to its pristine condition. Along the road to that ideal, however, there will surely be jobs that more modest nanomachines can fill.

One of these will probably be the removal of wrinkles. Wrinkles arise from many causes, and it is unlikely that any single remedy will cure them all. It should be feasible to design special nanomachines to attack specific symptoms, however, and it would certainly be possible to employ a number of different kinds of nanomachines simultaneously.

Most wrinkles occur in response to persistent folding, as with crows feet around the eyes and smile lines, and the deep nasolabial folds that develop on either side of the mouth. Part of the problem is that the skin has lost elasticity, although there may be any number of contributing factors.

A start at treating one of the basic symptoms would be a nanomachine designed to improve the skin's elasticity, or at least to prevent further deterioration. What gives youthful skin the ability to adapt gracefully to the body's contours and movements while remaining tough enough to resist the wear and tear of everyday life is its supply of collagen fibers. Old skin is softer and floppier than young skin; it has less collagen, either because its cells make less collagen than they used to or because they produce too much collagenase, an enzyme that destroys collagen fibers. Most likely, both processes are at work.

To attack this problem, a nanomachine could be designed to seek out the cells that produce collagenase and slow or stop them. Another and possibly simpler approach would be a nanomachine that attaches itself to collagenase in such a way as to jam its collagen-cutting mechanism. Either way, it would be necessary to ensure that the nanomachines would act only in the skin, where we want them to, and to monitor the process carefully to keep it under control.

One simple way to keep the nanomachines from attacking collagenase anywhere but in the skin would be to make them temperature sensitive. (We'll discuss how to do this later.) They could remain inert and circulating in the bloodstream until they encountered a high temperature, say 110° F (easily obtained locally at the surface with a hot pack). Then they would migrate out of whatever capillary they were in and go to work.

As with the previous products, a *feature* of such a nanomachine would be its inability to reproduce. This would permit control of the action by limiting the dose. But whereas for other applications the degree of change was a matter of vanity, in this case it would be a safeguard against removing so much collagenase as to make the skin leathery. Nonreproducing nanomachines also permit an incremental approach to the treatment. Instead of trying to prescribe the exact dose needed, the therapist could schedule a series of sessions in which no single application would contain enough of the product to do any harm.

Slender Now

A final example deals with what appears to be the number one beauty concern of American women: obesity. Here we venture beyond mere surface appearance and into bodily restructuring. And it must be admitted at the outset that not all kinds of obesity will yield to nanosurgical control. It appears that some obesity is genetic, some psychological, and some imaginary.

That said, however, there is still a vast market for a treatment that would short-circuit the ceaseless roller-coaster ride of conventional diet-splurge cycles. The ideal weight control product would allow a person to eat more or less what one wanted, continue to live a sedentary life, and maintain a figure that reasonably approximates one's ideal proportions. It should also be capable of spot application, to reduce a person's hips or abdomen for example, or wherever the problem appears to be.

At the same time, the product should induce a feeling of well-being, rewarding the user with positive sensations to make it easy to stick with the program and keep using the product. Just how that might be accomplished I leave as an exercise for the reader. As a hint, consider the role of endorphins, those natural mood-altering chemicals that your body is constantly ready to produce.

One possible approach would enlist the aid of the body's own immune system. A nanomachine could be designed to home in on a fat cell, plant a marker on its surface that falsely identifies it as a pathogen, and let the nearest phagocyte take over. An easy way to accomplish this would be to instruct the nanomachine to fabricate the marker on the spot by altering a surface protein so that the cell is no longer recognizable as "part of the family."

To make sure that such a product works only where needed (a person needs some fat storage capability, after all), it would probably be wise to make this nanomachine temperature sensitive too. Surplus fat generally lies just under the skin, and it is a notoriously good insulator, so it is easy to produce local temperature gradients across it. Any surplus nanomachines could just keep circulating until they were either eliminated naturally or put to use at some new problem elsewhere.

☐ Techniques and Strategies

Any protein product designed to circulate through the body and perform some specific task on a certain class of cells must fulfill several criteria. It must first of all conform to a rigid set of specifications imposed by the body's immune system. It must also have some way of recognizing the target cells and, in some cases, some way of "knowing" when it has arrived at an appropriate location in the body.

Foiling the Immune System

The need to get past the immune system's detectors is a daunting requirement in view of that system's perpetual vigilance and almost devilish versatility. As we have learned more about the system's functioning, however, small chinks have begun to appear in its armor. Presumably these chinks will widen with further research, of which we can expect a considerable amount in the ten to twenty years that are likely to elapse before the first nanosurgical products appear.

However, there is one surefire way to protect such a product from the immune system, even in our present state of ignorance: package it in something that your immune system instantly recognizes as completely familiar. And what could be more familiar than red blood corpuscles of your own blood type? After all, routine blood transfusions trigger no immune response.

A case might be made for using white blood cells, since these are naturally able to leave the bloodstream, crawl through tissue, and sense their environment. It might appear that hitchhiking nanomachines could simply wait, reading the macrophage's sensory input until it finds itself at its intended destination, and then slip out to do its work.

There are at least three major drawbacks to this scheme. In the first place, blood contains about 600 red cells for every white cell. Getting enough white cells to do the job would involve either taking and separating a large volume of blood or cultivating large numbers of white cells outside the body. Second, it seems quite unlikely that information about the white blood cell's environment travels to some central organ analogous to a brain, where a hitchhiking nanomachine could read it. More than likely, the cell's sensory equipment is confined to the tips of its probing tendrils and is designed to be simply oblivious to contact with "self" materials of whatever kind. In other words it reacts only when it senses something it recognizes as foreign, and then it would react in exactly the same way, wherever it happens to be in the body. Finally, it may be possible to slip a nanomachine past a macrophage's defenses and into the cell at the outset, say by inactivating the white cell with cold. It is hard, however, to imagine how an emerging nanomachine could escape instant destruction by the fully active macrophage it had been riding. It would be much like the case of the smiling young woman from Niger, who went for a ride on a tiger.

On the other hand, slipping nanomachines into red blood cells would be easy and completely harmless, assuming the product is correctly designed to be nontoxic and to penetrate the cell membrane without damaging the cell, as viruses do now. A sample of blood of the correct blood type would be separated into its components by centrifuge or otherwise. With no white cells to interfere, the nanomachines could infiltrate the red blood cells upon contact. A rinse to remove stragglers that failed to make contact would probably be in order before injection.

Packaging the nanomachines inside red blood cells gets them past the immune system, but it isolates them from many of the clues that might be used to activate them. Any attempt to poke a molecular fragment complex enough to act as a sensor out through the cell wall would invite attack by the immune system. What to do?

One possible approach would be to take advantage of the red blood cell's internal environment. All of the nanomachines we have considered are supposed to do their work at or near the body's surface. And that is where blood temperature is lowest and where the hemoglobin in red blood cell's becomes most oxygen-depleted. The nanomachines should be able to recognize these environmental clues while still inside the red blood cells.

Once the nanomachines get the "jump signal" that tells them they are in their "drop zone," they emerge from the red blood cell and begin to search for their programmed target cell, which they can recognize by its surface proteins. Some nanomachines may fall prey to wandering phagocytes, but most of them, having been released close to their targets, should succeed. Those that do will be safe from phagocytes as soon as they slip inside their target cells.

Sensory Equipment

So far we have postulated nanomachine molecules able to act on environmental cues such as temperature, oxygen depletion, and chemical concentration. The question is, how do we make molecular machines with these sensitivities? The answer is, with sensing mechanisms much like those found in natural molecules.

By now we have discovered many examples of proteins (enzymes) that bind to a certain shape of molecule, often with exquisite selectivity. The common feature of all these examples is a receptor site on one molecule that closely matches a shape on the other molecule. In most such instances, once the two molecules have come together they tend to stick. In other words, it takes only one molecule to fill the site and trigger a given reaction.

To make a sensor that responds only above a certain concentration of the target chemical, we need something more. We need to give the nanomachine two weakly binding receptors linked in such a way that both must be occupied before the nanomachine responds. That way nothing happens until the concentration is high enough to provide a reasonable chance that the second site will be filled before the molecule occupying the first one shakes loose. It would be possible to control the concentration required to trigger the device by adjusting the binding energy, and thereby the residence time, in the receptor sites.

For a sensor to detect above-normal temperatures, we need something similar but opposite. In this case we start with the active site already filled with a target molecule in such a way as to jam the nanomachine's mechanism. However, we design the attraction between the two so that they break apart under molecular bombardment above a certain temperature, freeing the nanomachine.

For a below-normal temperature sensor, we start with a nanomachine consisting of target and receptor molecules already connected by a flexible hinge, but with enough mismatch between their shapes that they are only weakly attracted at normal temperatures. As the temperature decreases, the two halves of this hinged ensemble spend more and more time in contact. Eventually, they stay together long enough for a slow-acting latch to function, holding them together permanently.

When a predetermined sequence of triggers fires, the nanomachine is activated to perform its intended function. We can envision the activated nanomachine unfolding itself into the required shape like one of those "transformer" toys, an obedient robot ready to go into action.

Intercommunications

For most of these applications, each nanomachine would be on its own, carrying out its instructions on atoms and molecules within reach of its manipulator arm or within a single target cell. For control over hair waviness (which requires that the nanomachines affect the hair's cross-sectional shape), however, hundreds of nanomachines in a particular follicle must work in concert. This means that they will have to communicate, at least in a rudimentary way.

Since such traits as hair color and hair waviness are clearly inherited, it seems reasonable that the eventual channel for this communication would be genetic—by manipulating a person's DNA. But long before we have nanomachines capable of such fine control, we should be able to find ways to coordinate less-sophisticated groups of them to override the genetic instructions.

One such crude signaling method for hair waviness could be by means of a chemical gradient. The first nanomachine to gain a foothold in a particular follicle would stimulate its cell to generate a chemical marker, thereby staking its claim as the controlling nanomachine. This chemical marker would inhibit each subsequent nanomachine from emitting the same signal, thereby establishing a chemical gradient from one side of the hair follicle to the other. From this, each nanomachine could estimate its position with respect to the "master" nanomachine and modify its cell's hair-making activity accordingly.

☐ **Licensing**

It should be clear that none of the techniques so far described, with the possible exception of the depilatory and tooth-protector nanomachines, could in any way be marketed over the counter. Each of the injectable products involves at least some technical sophistication and probably would need to be administered by medical professionals. What will be the legal ramifications?

In the first place, as injectable pharmaceuticals, each of these products would come under the licensing authority of the Food and Drug Administration. (I think we can assume the continued jurisdiction of the FDA. This agency has, on the whole, done yeoman's service in protecting the American public from hazardous nostrums, despite recent criticism of its slowness in approving untried methods of controlling AIDS. And even totally useless government agencies have amazing powers of self-preservation.)

FDA approval is a long and costly process, involving extensive testing for both safety and efficacy. This process would be undertaken primarily by drug companies, not cosmetic manufacturers, since externally applied cosmetics need to be certified only for safety.

At present, the preliminary testing of any such new technology would be done on animals, although there are many opposed to this practice. And there are valid arguments against animal tests in this case. It would not be easy to prove, for example, that a nanomachine designed to work in mice or rabbits would necessarily perform properly in humans.

At the same time, the idea of testing a brand new technology on humans is far from attractive and involves serious ethical questions. A possible way out may be to develop surrogate human tissues on which to perform these preliminary tests. Tissue culture methods so far developed, however, work only on single cell types. They are incapable, without considerable development, of producing the complex structures of even such a relatively simple organ as skin.

We would have no problem, of course, if we had self-reproducing cell-repair nanomachines to work with. They could monitor the internal workings of each cell in a growing population, reproducing in synchrony with the cell they inhabit, and guide the development of its descendants along predetermined pathways to produce any desired tissue component. Just for the testing of cosmetics alone, such technology could grow acres of human skin,[4] complete with nerve endings, capillaries, hair follicles, sweat glands, and even subcutaneous musculature and fat cells—thriving under glass in racks of oversized Petri dishes in a laboratory. (It would also be important to cultivate mucous membranes, since these are much more sensitive to many irritants than the outer skin.) Incidentally, this system could also be used to grow all sorts of replacement organs for transplantation, relieving the chronic shortage. But by the time such capabilities matured, organ transplants are likely to be unnecessary except for the treatment of severe, life-threatening injuries. In-body repair, using injected nanosurgical nanomachines, would be far preferable in every way whenever there was time for it to work, and it could be used to treat conditions we now tolerate until they become emergencies.

Long before cell-repair nanomachines become available, however, a much simpler solution will have presented itself. Recent experiments show great progress in grafting human tissue onto animals without provoking an immune reaction. Carried just a bit further, this xenograft technique should enable us to produce living surrogate organisms whose tissues mimic those of humans closely enough for reliable test purposes but with just enough of a nervous system to control such vital functions as respiration, digestion, and circulation. Naturally, they would have no sensory equipment that would allow them to feel pain. It's hard to see how animal-rights activists could object to laboratory tests on living nonanimals such as a pillow-shaped object covered with human skin.

☐ Economics

What will be the likely response of the cosmetic industry? Probably not much—at first. As usual, the first feeble attempts will be greeted with derision and ascribed to the work of crackpots and charlatans. As the procedures became more mature and routine, however, and more affordable, the established companies might be forced to choose between opposing the new technology or buying in to it.

Some of the largest of them will probably hedge their bets by doing both. After all, Chevrolet competes as fiercely against Pontiac as it does against Ford. It will be interesting to watch the media battles that will predictably ensue.

☐ Distant Prospects

I have tried so far to touch on only the easily foreseeable and immediate consequences of the simplest nanosurgical nanomachines, but it is also worthwhile to sketch more distant possibilities. Once fully capable cell-repair nanomachines hit the market, there will be little the imagination can conjure up that would be impossible. Permanent rejuvenation of face and body is only the most obvious result.

Nonsurgical Nanomachine-based Cosmetics

In the end, however, it is unlikely that cosmetic nanosurgery will completely supplant conventional cosmetics. After all, there is something to be said for being able to adjust one's makeup to one's mood or costume, and to change back and forth between different looks quickly. And it should be noted that the primary consumers of beauty preparations today are not the elderly, who might presumably need them most, but the already young and beautiful.

Instead, it is likely that nanomachine techniques will take over the task of manufacturing cosmetics, a job for which programmable, self-reproducing nanomachines are ideally suited. The ability to manipulate individual atoms and promote or inhibit specific reactions could lead to all sorts of new fragrances and materials tailored to the cosmetic industry's needs.

Some cosmetics, in fact, might consist of nanomachines. Who could resist a fingernail polish or eye shadow, for example, made of units that automatically aligned and spaced themselves to produce a diffraction grating? Rainbow colors would flash from such a product, depending on the lighting and on the angle from which it was viewed. The colors might also be made to change with skin temperature or other environmental cues.

The Total Makeover

Eventually, under the care of a competent practitioner backed up by talented molecular programmers and a responsive pharmaceutical industry, it should become possible to mold the face and body to whatever shape might be desired. Each person who cared to could achieve his or her own ideal of physical perfection or, for that matter, whatever frightening or gruesome effect they wanted.

Inevitably, there will be people who don't know how to leave well enough alone. Many who never liked their own youthful appearance will opt instead to copy some popular model or other sex symbol. It could become very confusing, with dozens of pop-idol look-alikes crowding the parks and boulevards of our future metropolis. Some may not relish the prospect, but we may never see the last of the Elvis clones.

■ 4 Diamond Teeth

Edward M. Reifman

The tooth is out; once more again the throbbing, jumping nerves are stilled. Reader, would you avoid this pain? Then have your crumbling teeth well filled.
—David Bates

Diamonds are Forever.
—Ian Fleming

☐ Losing Teeth

Recent, exciting strides in permanently attaching false teeth to the jaw, or using "space-age" plastics to restore an old tooth to its natural strength and beauty, can make it easy to forget that, not too long ago, millions of people in the United States had *all* their teeth out by the age of twenty-five or thirty in order to put an end to bothersome, rotting teeth. In fact, through the 1950s, families that could afford a dowry would often provide for the purchase of a spouse's dentures. Today it seems that everyone over the age of fifty is anxious, to a degree, about the prospect of losing some or all of their teeth.

Is tooth loss inevitable? No. When nanotechnology arrives—two or three decades hence—it will arrest the genetics behind tooth loss and the accompanying thinning and deterioration of the jaw bone. Moreover, nanotechnology will be able to reverse this process of degradation. Let's look at some of the latest advances in the field of dentistry and then consider how a visit to the dentist will be irrevocably altered when nanotechnology becomes a reality.

☐ Saving and Replacing Teeth in the 1990s

Dental implants, or permanently attached false teeth, have become very popular. Metal anchoring units surgically implanted into a patient's jaw allow the bone to flow around and bond to them. This bonding process, known as *osseointegration,* can take up to six months. Once complete, a dentist attaches one or more artificial teeth to these strong anchors. After several visits to the dentist over a period of several months, the satisfied patient with a brand-new smile is able to chew on an apple the way he or she did when young and, well, full of teeth.

Unfortunately, malpractice suits related to this practice have increased, partly because the implant process is sometimes carried out too swiftly or the implants are improperly fitted. In some cases, the procedure should never have been recommended. With aging, a patient's jaws can become so weak or diseased that implants cannot find a firm foundation; implant technology available today does not replace the jaw bone or mitigate bone thinning as we get older.

Dentists have recently begun to use computer-aided design and manufacturing (CAD-CAM) systems to fabricate porcelain caps. With one end of a fiber-optic light wand plugged into a desktop computer, the dentist moves the free end over the tooth. Instantaneous measurements of the tooth contours (before and after the tooth is drilled) are transmitted via the computer to a portable milling machine, where a tooth-colored cap is fabricated. It is then permanently cemented onto the prepared tooth—all within 90 minutes. No more suffering for two weeks with an ill-fitting, temporary tooth covering, all the while dreading another trip to the dentist for the final procedure.

A genetic approach, only now on the "biological drawing board," could make even this CAD-CAM procedure obsolete. By the turn of the century, your friendly neighborhood dentist will routinely repair that cracked molar, not with computer-designed caps, but simply by adding a tooth-colored "paste"—derived from cloned enamel genes—to almost any broken-down tooth. Thanks to a few small alterations in the DNA, this biocompatible material could be stronger than the original enamel and more resistant to cavities.

☐ Dental Care in the Nanofuture

Let's get into a dental time-machine set for the year 2020 and visit the posh dental suite of Dr. Harvey Smile-Maker. His office is located, naturally enough, in Southern California, where so many new dental techniques are first attempted—for better or worse.

Dr. Smile-Maker peers into the oral cavity of his new patient, Mr. John Garbage-Mouth. Mr. Garbage-Mouth's teeth are in a state of chaos. He steadfastly refuses to use his robotic tooth-flosser before retiring for the evening. His habit of eating chocolates has accelerated an already severe case of tooth rot. The last straw was when his horrific mouth odor finally drove his girlfriend to leave him. The man is desperate for (among other things) some good dental care.

Mr. Garbage-mouth timidly reclines in an air-cushioned dental chair that is computerized to custom fit the contours of his body. As our patient hopefully nods at the holographic "We Cater to Cowards" sign, the dentist adjusts a handheld, portable positron-emission tomographic (PET) machine. The PET scan will show the precise, three-dimensional characteristics of the toothless regions of his mouth and will summarize the following: normal/abnormal bone and gum densities, all vessels, and the specific sites where further tooth or jawbone loss is most likely to occur. All this information is used in conjunction with other scanning devices (still on the drawing boards in the 1990s) that feed the data directly into a computer. Powerful "expert system" software almost instantly determines optimal jawbone sites and the precise amount of biological ingredients to be used for tooth rehabilitation.

The treatment begins. Patches attached to the patient's gums send out electrical signals that deaden the appropriate gum and jaw nerves within seconds. (Injections of anesthetic will be a thing of the past). A computerized robotic arm designed to work inside the mouth begins drilling tiny, cylindrical holes into the jawbone at positions selected by the PET scan. Then the robotic arm grasps a syringe containing a few drops of "seed" material containing uncountable programmed assembler molecules and injects them into the drilled sites. A dental impression tray containing the construction materials necessary for building the tooth is carefully fitted over the site.

Let me discuss the tray for a moment, for it is the centerpiece of this fantastic technology. It resembles a disposable plastic dental-impression tray of the 1990s that was used to make a mold of the prepared tooth so that the dental lab could fabricate a cap. On closer inspection, we see the interior or backbone of the tray is lined with trillions of molecular-sized computers. Working in parallel, they precisely regulate the flow of construction materials—mixtures of artificial calcium hydroxy-apatite crystals (the molecules that bone is made of) and natural and semisynthetic organic molecules, as well as other molecules never seen before—as they come in contact with the seed material within the jawbone sites. The process is initiated and powered by an excimer laser pulsing through a fiber-optic bundle. Thus excited, the nanoseeds configure the scaffolding for a new tooth, consuming the provided construction materials flowing from the tray.

"Bleep, blurp, blam!" Before you can say "dental floss," a beautiful, perfectly shaped tooth begins to take form. And don't worry about long dental appointments: "nanodentology" will produce pearly whites within an easy morning appointment. Or just about long enough for our patient to get those chocolate cravings again.

This wonderfully high-tech molar will duplicate or improve upon the clinical and morphological characteristics of the original, unworn tooth that existed prior to decay and breakdown. Unlike the cloned-gel material discussed earlier, there will never be the danger of tooth fracture or decay at the interface of natural and artificial tooth substance. The entire replacement tooth, root and crown, will consist of a single, solid, cavity-proof matrix.

Constructed with atomic precision, the new tooth will be a substantial improvement—considering strength and durability—over mother nature. Biting an apple will feel as comfortable and satisfying as if you had the teeth of a teenager. What a far cry this dental era will be from the 1990s when it took stressful surgery, discomfort, and a wait of up to nine months for an implanted artificial tooth to achieve a similarly strong bite.

☐ Diamond Jaws

Nanotechnology will deliver the holy grail of dentistry: long-lasting, cavity-free teeth. But just a few months or years later, *advanced* nanotechnology will deliver another coup: arresting or neutralizing the genetics behind a degenerating, aging jawline. For if we are able to fabricate perfect, ageless teeth, could not the same technique be used to augment and strengthen a thinning mandible? And what material would be the strongest and most durable to use for this procedure? Diamond.

Our patient could be an eighty-year-old with few or no teeth and a jaw that the ravages of time had narrowed to only the thickness of a pencil. In this procedure, uncountable assembler molecules from the dental tray are quickly depositing a diamond substructure that fills in and augments the disease-deficient and thinning areas of the patient's jaw. At the conclusion of this treatment, the patient could well end up with the youthful jaws of a twenty-year-old—or better. Forever freed from the ravages of a deteriorating jawbone, not to mention bone cancers, numerous bone disorders and other distasteful age-related phenomena, we could enjoy the jaws, teeth, and smile of youth, perhaps for decades.

We could eventually see the replacement of the entire jaw and teeth with a diamondoid matrix. But why stop there? We can expand this approach to improve or replace the body's entire skeletal structure. We've seen Batman, Superman, even Lawnmower Man—look out now for Diamond Man. Only this amazing human will be each of us, an Everyman, finally freed from bad teeth, worn-out jaws, and other bone maladies.

What an era the human race is fast approaching! A time when crutches, bone casts, and dentures will exist only in museums. And if you decided to never brush or floss again, the worst you might get is a case of diamond breath. To help you hold out for the nanotech miracles of your local dentist in a couple decades hence, floss and brush well. The world of dentistry is about to offer exciting improvements—all for a healthier you.

 **II   Mostly Outside**

Harry Chesley

First you see video. Then you wear video.
Then you eat video. Then you be video.
—Pat Cadigan

☐ **An Opening Selection**

So what good are nanomachines? How will they have an impact on *my* life? What will we use them for? How pervasive will they be? Consider these possible applications:

- *Full-wall video screens* for television and video games.
- *Full-wall cellular automata* for chaotic and fractal entertainment.
- *Full-wall speakers* for people without neighbors.
- *Programmable paint* that changes color, texture, and pattern on command.
- *Reprogrammable books* that retain the tactile feel of today's books, and the ability to keep your place by thumb, but which have changeable content.[1]
- *Self-adjusting contour chairs* that change shape to match the person sitting in them.
- *Paint-on board games* that can be applied to any surface; they find the edges and scale the board to fit in the space available.
- *Board games with billions of moving parts,* allowing economic, logistical, and military games with incredible depth of simulation.
- *Windows with variable transparency* that allow you to select how transparent or opaque they are, allowing you to maintain a particular interior light level—their transparency adjusts dynamically as the outside light level changes.

- *Walls with variable transparency,* to take the concept to its logical conclusion.
- *Retractable walls and ceilings,* to let you dynamically change your apartment from small, cozy, and soundproof to big, open, and essentially transparent.
- *Walk-through walls* composed of millions of small, spring-loaded hinges that allow it to break apart harmlessly when you walk into it, and then snap back together again; can be embarrassing—and dangerous—when locked.
- *Programmable rooms* with configurable walls, ceilings, floors, and optics; giving rise to the two room house: the other room reconfigures to whatever you want next.
- *Seamless doors, cabinets, and closets* with no handles or hinges or cracks—but hopefully outside ventilation.
- *Odor-eaters* that suck room air through themselves and filter out dust and odor particles.
- *Ever-sharp knives* composed of nanomachines that cooperate to rearrange themselves so the knife always has a perfect edge.
- *Reversible shrink-wrap bags* that seal and unseal on command.
- *Always clean, nonslip bathtubs* that are smooth at a molecular scale, so that dirt doesn't stick, but which indent so you never slip.
- *People-scrubbers* that run off your body heat and constantly scrub dirt and odors from your body, so you never need to take a shower.
- *Active food* that squirms around as you eat it, but becomes inert once swallowed—very popular with kids.
- *New techniques for terrorism and assassination,* such as food that, once swallowed, turns into razor blades.
- *Solar-powered migratory houses* that slowly move themselves over land; it may take your house a year to get to Florida, but you get to live in a different neighborhood every week.
- *Zipperless pants* that come apart on command—yes, that does give rise to many practical jokes.
- *Spray-on pants,* for real.
- *Temperature-sensitive cloth* that changes the tightness of the weave depending upon the ambient temperature.

- *Mood- and context-sensitive clothing and jewelry* that change colors and patterns depending upon the local sound and light levels, the wearer's heart-rate, body temperature, and so on—just in case disco comes back into style.
- *Paint-on watches and thermometers.*
- *All-emergency clothing* that expands to several feet in thickness and becomes gently resistant to deceleration (like an air bag), inflates with air (like a life vest), radios your position and vital signs, and activates on command or under sudden deceleration—useful in car crashes and earthquakes, or when falling in the ocean or off a cliff.
- *Temperature insensitive gloves* with receptors that conduct pressure and texture but not heat or cold.
- *Active-bounce shoes* that use energy from your photocell shirt to "push off" with each step; or, if you want exercise, make it feel like you're constantly walking uphill.
- *Tool-less construction material* that breaks apart along a perfect seam on command—and glues together seamlessly.
- *Solar-powered dirt movers* that relocate hills, one grain of dirt at a time.
- *Time-release and time-retract medicine* that exposes a chemical catalyst at a set time, and then retracts it at another time.
- *Ditto for recreational drugs.*
- *Ant poison* that tastes good so ants take it back to the nest, where it expands to several times its original size and density.
- *Whole-building shock absorbers* for those living in California.
- *Modeling clay* that works with special tools to make perfectly straight cuts, add textures, change colors, and so on; and converts itself into a computer-aided design (CAD) specification on command.
- *Instantaneous scale models* from CAD designs.
- *CAD-driven, self-constructing, self-destructing molds* for building things out of cements and plastics.
- *Paint-on, computer-readable white-boards* that can be sprayed onto any surface and, when written on with a special stylus, read by a computer.

- *Self-testing construction material* that constantly checks its structural integrity and current load, emitting an audible sound when unsafe.
- *Industrial espionage*—well, you can just imagine.
- *Floorless elevators* in which you just step into space: it "catches" you part way down and smoothly decelerates you—will probably take a little getting used to.
- *Lateral-force sensitive tires* that change shape as you take a turn.
- *Bulk material shock absorbers* that work like normal shock absorbers but have no macro-scale moving parts.
- *All-around bumpers* that extend themselves if they sense someone getting too close and unbend themselves after impact.
- *The ultimate air bag* that collapses at an optimal rate to minimize damage.
- *Safe deceleration from any speed,* by combining all-around bumpers and ultimate air bags—giving rise to 90 MPH bumper cars as a popular sport among teenagers.
- *Extensible cars* that extend on command from two- to four- to six-seaters.
- *Stationary moving walkways* that use microcilia or microrollers to move passengers, but which are themselves stationary.
- *Computer-generated holograms* that require huge amounts of easily available computing power.
- *Contact lens virtual reality* consisting of an arbitrary mix of the real world and a superimposed computer-generated virtual world.
- *Fully encased virtual reality users* where the virtual reality environment completely surrounds the user, providing full-bandwidth, visual, auditory, and tactile stimulation and "realistic" physical feedback.

☐ **Changing Lives**

Some new technologies change every aspect of our lives. By this I don't mean a change in our ability to do some one thing, or a small collection of things, that we were unable to do before. I mean change that pervades every part of our day-to-day lives—change that really alters the way we live.

Electronics is probably the best known contemporary example of this sort of change. Electronic machines have changed the way we work, play, and live—or don't live. They have created whole new industries, including the concept of a "knowledge worker."[2] The information industry, one of the hallmarks of modern civilization, could not have grown to its current size and complexity without electronics. There would be no computers without electronics— Charles Babbage notwithstanding. There would still be entertainment, of course, but there would be no radio or television or stereo. (I'll leave it to the reader to decide if that's good or bad.)

Nanotechnology is a technology destined to change every aspect of our lives. It will allow marvelous new machines and applications, things we can only just imagine, and things we can't imagine at all. It will have an overwhelming impact on our daily lives. And it will happen sooner than you think.

This chapter will show how pervasive an impact nanotechnology will have on your life—and that it won't take wild, far-fetched concepts or Herculean feats of engineering either. Fairly simple, easily engineered nanomachines are capable of performing tasks that have value—or at least impact—at virtually every level of our lives. We don't need to develop full-blown replicating assemblers to enter an entirely new world.

It may be that what is described in this chapter is all that can be predicted. As others have pointed out, once self-enhancing artificial intelligence (AI) has been built, we quickly snowball into a sort of singularity.[3] It is very difficult to predict what entities much more intelligent than you or I will create, or for what purposes they will use the things they do create. Artificial intelligences that can redesign and improve themselves will develop at exponential rates, and it is quite possible that such intelligences will appear shortly after nanotechnology creates molecular-scale computing devices, devices with truly awesome power.

Indeed, we may not survive the process. This sort of AI is not only unpredictable in its capabilities, it's unpredictable in its intent. Our only hope for survival may well be to get there first by creating enhanced human beings with equally high levels of intelligence. But, by definition, such intelligence amplification (IA) will produce enhanced humans that are only partly human—and the human part will not be the greatest part. Again it becomes difficult to predict what such entities will want from their technology. And it is hard to predict what they will do with nanotechnology.

This isn't meant to sound pessimistic. I'm looking forward to being one of the enhanced humans leading the way into new territory. It will be very exciting. But one of the reasons for excitement is the fact that it's unpredictable.

In the meantime—and it may take several decades—we'll have the technology, machines, and applications described in this chapter to keep us busy. View this chapter as an interim prediction: it aims to cover the territory from the first realization of nanotechnology to the creation of true, inexpensive replicators. After that, it's wide open. You may find glimpses of the future here and in the other chapters of this book. Or you may find that we're in the same position as a prehuman ape trying to imagine what electrons are good for. We certainly don't have the perspective, and we may not even have the intelligence or imagination—yet.

☐ Nanomachinery

There are usually several ways to engineer a device, nano-scale or macro-scale. The final design is based on hundreds if not thousands of practicalities, efficiencies, availabilities, and experiences. Since many of these are currently unknown for nanotechnology, this chapter presents a plausible rather than certain design for a general purpose nanomachine that can be programmed for several different uses.[4] The following paragraphs present a description of the scale, shape, processing power, communication systems, energy requirements, and construction techniques of one such nanomachine.

Scale

Although the active elements of this device are measured in nano-meters, the machine itself measures approximately a micron or a millionth of a meter. There are several reasons for this. The machine needs to be resistant to damage and capable of affecting its environment without breaking up. In addition, the device needs a reasonable amount of computing power to accomplish its tasks, and computational machinery, especially memory, requires a certain amount of space. The device also needs to influence the outside world, physically pushing against things or reflecting visible wavelengths of light. Finally, the energy requirements of many applications will require power supplied from outside the device, either electrically or mechanically. Such an interface requires a certain minimum size.

Shape

Spheres are the most efficient shape, in terms of volume-to-surface-area, and the most efficient and stable spherical approximations are the various geodesic forms (figure 5.1) studied by Buckminster Fuller.[5] A micron-scale geodesic structure built with carbon rod struts would form an extremely secure exoskeleton for housing nanomachinery. Whether such forms are based on the popular soccer ball structure of buckminsterfullerene or other geodesics is a matter of

 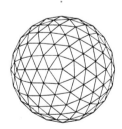 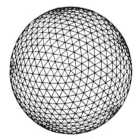

Figure 5.1 Regular geodesic two-, four-, and nine-frequency icosahedrons. (Reprinted with the permission of Simon & Schuster, Inc., from *Synergetics* by R. Buckminster Fuller. Copyright © 1975 by R. Buckminster Fuller.)

design. In any case, a series of structurally linked, concentric geodesics could easily form a secure shell for nanomechanisms, providing them with structures to push against.

A one-micron sphere provides a volume of about 520,000,000 cubic nanometers. Dedicating a third of that to structural supports—a conservative estimate—leaves just over 340,000,000 cubic nanometers for functional machinery.

By extending arms through the vertices of the geodesic, the machine can influence and communicate with the outside world. Simple collapsible rods allow external objects to be pushed away and can also make electrical power contacts. Rods that end in protrusions—perhaps dynamically extensible themselves—allow other nanomachines to grab the rod. In this way, two nanomachines can both push and pull each other and establish secure power and communications linkages.

To give each rod a good reach, we can use a telescoping design. To allow them to fit snugly within the structural framework, each rod, when collapsed is a little over a third of the machine's total diameter. We then need nine telescoping components to give the rod a reach of three microns, or three times the diameter of the machine itself. If each rod component is a hollow cylinder with a wall thickness of ten nanometers, the entire assembly will fit in a cylinder 180 nanometers in diameter and 400 nanometers in length, and occupy about 10,000,000 cubic nanometers.

If the basic design is icosahedral, and one rod extends through each primary vertex, twelve rods are required, occupying a total of 120,000,000 cubic nanometers.

Processing Power

Each machine needs storage and computing facilities. It seems that it will be possible to build a 1,000 MIPS (million instructions per second) molecular computer that fits inside a cube 0.4 microns on a side (i.e., in a volume of 65,000,000 cubic nanometers).[6] This is roughly 1,000 times the computing power of today's personal computers.

In addition to the processor, we need a large memory, primarily to store the programs that deal with the situations the machine may encounter as it operates in the real world. Efficient memory systems should allow for the storage of one bit per 5 cubic nanometers. Assigning 50,000,000 cubic nanometers of our remaining space to storage can provide 10 megabytes, including room for error correcting codes.

Computing power requires energy—often a scarce commodity. For applications described in this chapter, we don't need anything close to 1,000 MIPS for each micron-sized device; one personal computer equivalent is more than enough for our purposes. This reduces the energy needs a thousandfold. Molecular memory does not require large amounts of energy. Indeed, it only requires energy when it's being accessed. By reducing our requirements for computing speed, we also reduce our needs for space and energy.

Communication

These devices need to communicate with each other and with human users. In fact, both types of communication will be needed simultaneously as the machines interact with each other and the users to carry out the user's wishes.

For many applications, a number of nanomachines must work cooperatively toward a single goal. To do so requires communication between the devices. In some circumstances, it may also be desirable to employ large amounts of computing power. By cooperating, the machines can turn their millions of individual processors into a single, much more powerful, multiprocessor computer.

Peer-to-peer communication is achieved most easily by electrical connectivity. Even a fairly sparse set of connections can provide global communication if the individual machines forward messages to other machines.

User communication is a different problem. In many applications—though not all—the user must tell the machinery what to do. This can be accomplished via optical communications, using the same technology that is used today for remote control of stereo

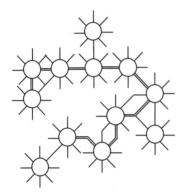

Figure 5.2 Network of nanomachines with communication paths shown.

components. Surface nanomachines detect the optical signals, interpret them, and pass them on to machines not in the line of sight of the control device.

Communications between machines is actually a computational issue. It uses the computing power described in the previous section and does not require any additional space. As shown in figure 5.2, communication channels can piggyback on structural and power components of the machinery so that no additional space is required either. However, we will need to allocate some 10,000,000 cubic nanometers per device for signal detection and amplification.

Energy

Some nanomachines can be designed to operate passively. But many will require power to actively affect the environment. In fact, this is probably the most limiting aspect of the nanomachine design problem, and one often overlooked when uses for nanomachines are considered. Although nanotechnology provides many solutions for mechanical problems, it provides few solutions for energy production—and nanomachines can be very energy thirsty.

In addition, heat dissipation can be a seriously limiting factor in the design of nanomachines. It can restrict the density of the ma-

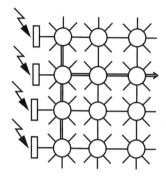

Figure 5.3 Machines with photocells routing power to other, interior machines.

chinery, or it can increase the size of the machines by requiring the addition of channels for heat removal.

For applications requiring large amounts of power, where substantial physical work must be done, direct connection to an external electrical or mechanical power source is essential. If all coherently functioning nanomachines are physically adjacent, connecting some will allow them to route the power to the others through electrical junctures or mechanical couplings or both.

For applications requiring small amounts of power, photocells on surface nanomachines may suffice. As shown in figure 5.3, surface machines would convert available light into electrical power and then route the electrical power to other machines as needed.

For applications requiring only occasional bursts of power, or operating in environments with a great deal of physical motion, simple physical shaking may be sufficient to supply the needed power—like an old, mechanical, self-winding watch. In some cases, the user may literally shake the nanomachines before using them. In other cases—in an automobile or on clothing—normal environmental movements may supply enough energy.

Wherever the power comes from, and in whatever form, a mechanism is needed to store it and convert it to a usable form. About 15 percent of our nanomachine, or 75,000,000 cubic nanometers, will be reserved for power storage and conversion.

Construction

As described, these nanomachines are very simple. They are composed of simple, nested geodesic skeletons and extensible rods, with no more computing power than today's personal computers. It is not such a tremendous task to design machines with nanometer-scale components, but how can we construct them?

At first, the answer will no doubt be, "One at a time." As of this writing, we can manipulate individual atoms at the nanometer scale and build more complex machinery at the micron scale (e.g., computer chips). Given the pace of technological development, we will certainly see machines of the complexity described here built "by hand" fairly soon. But to be useful, we will need trillions of them— one trillion would fill a single cubic centimeter.

In time, limited-scale mass production will arrive. Shortly after the first effective assembler has been built by hand, they will be manufactured in quantity—and used in turn to manufacture the machines described here. If a factory has ten million assemblers (which might have taken years to create), each capable of producing ten thousand nanomachines a day, it will take two weeks to produce a cubic centimeter of finished goods. During this period, nanomachine-based products will be rare and very expensive.

Eventually, replicators will be used to produce assemblers, which will then produce nanomachines for final use (figure 5.4). During this period, replicator technology will be viable only within very constrained environments. I expect that this period will last for some time, since designing unconstrained replicators is a very hard problem. Although this period will not allow consumers access to replicator technology, it will be a time filled with inexpensive, easily accessible nanomachinery of the type described in this chapter.

These three stages of nanomachine production seem most likely, but there are other possibilities. A crystal array might simultaneously "stamp out" millions of nanomachines, using the same techniques we use today to manipulate individual atoms. But this technique would likely produce only very low yields.

We may also use genetic engineering to adapt and exploit the nanomachinery of living organisms to create nanomachines similar to

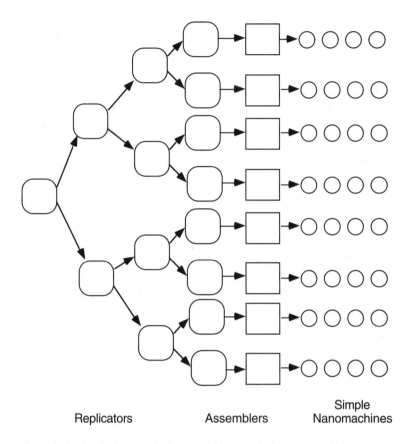

| Replicators | Assemblers | Simple Nanomachines |

Figure 5.4   Replicators producing assemblers producing nanomachines.

the ones described in this chapter. But this would lead to machines
with significant restrictions, depending upon the capabilities of the
organisms used.

☐ **Material Properties**

Having described individual nanomachines, and how we might
manufacture them, we can now consider how to use them to construct materials with interesting properties. Here we'll consider the
properties of bulk, solidity, shape, optics, and force.

Bulk

The first, and most obvious, aspect of a material is its bulk: its size and density. For humans, a single micron-scale machine is invisible and ineffective. It takes trillions of them to make something we would notice. Once we have enough machines to make a material of useful size, we can change its density simply by extending and retracting the nanomachine arms described earlier. With arms three times the diameter of the body, we can effect 4-to-1 changes in density. If the arms can directly oppose each other, we could double that distance. But this would require much more substantial arms. Assuming a conservative design, the changeable density ratio is limited to 4 to 1.

Changing bulk takes energy. In the process of extending the arms, adjacent machines may need to be pushed away. If a good power source is immediately available, there's no problem, but power will often be in limited supply. Still, there are several possible solutions. By employing a ratchet mechanism, arms can be constrained to extend (or contract), so that density can be changed by shaking or "fluffing" the material. In situations where there is a lot of physical movement—a car's shock absorbers, for example—this is even easier. Or, if we're not in a hurry, low-level, external power sources such as sunlight may suffice.

Solidity

Viscosity is a measure of how much internal friction a liquid has, or how easily it flows, or, in a sense, how *liquid* the liquid is. We can adjust the viscosity of our machines by having them grab—or not grab—onto each other. In the extreme cases, we end up with a solid, where all the machines are grabbing each other firmly, or a fluid in which all the machines tumble past each other without constraint. Between these two extremes, we can set the granularity of the material and independently set its resistance to flow.

In addition, we can also determine the strength of a solid. If all the machines are initially set to grab each other, they can be programmed to release as certain levels and types of forces are applied. We can make a very hard solid, or a crumbly one, or one that's very resistant to horizontal force but not at all resistant to vertical force. We could also make solids that are extremely solid at low pressures, but which give way at slightly higher pressures. Or the other way around.

Shape

Shape has two components: overall form and surface detail. Object form is the shape of the object itself—a cube or sphere or lawn chair or teapot. Surface detail determines the texture of the surface, the smoothness—or sharpness—of the edges, and so forth.

Nanomachines can coordinate with each other to build a particular shape, or, with much less expenditure of energy, they can let users set the shape. For example, a special tool could be used to indicate a plane along which to break. All machines along that plane would simply let go of their neighbors, so that the material breaks apart along a perfectly straight cut. The process can also be reversed to seamlessly recombine two pieces of material. Using techniques like this, arbitrarily complex designs can be fashioned without requiring the material to consume large amounts of energy or to solve complex global communication and coordination problems.

Texture, forming and maintaining desired edges, and otherwise coordinating the surface structure requires an awareness among the nanomachines of the existence of the surface. Those machines at the surface can sense that fact because they are only in contact with other machines on one side. By communicating with other adjacent machines, they can determine the current shape of the surface and what needs to be done to reach the desired shape. As always, it may require some energy input to effect the change. A nanoengineered knife, for example, may need to be tapped against the counter a few times to supply the energy required to resharpen it.

Optical Properties

There are two main aspects to optics: changing the machine-level characteristics of a material and determining what this looks like to a human.

Micron-scale machines are large enough to reflect visible light. This is both a blessing and a curse. On the one hand, it's fairly easy to change their reflectivity characteristics; on the other, transparency becomes a challenge. While complete transparency seems quite hard, a variety of optical effects could be derived by mixing the machines with inert, transparent tiles that the nanomachines could rearrange—covering them or causing them to overlap.

With this control, we can select an object's overall appearance. The easiest effect to achieve is changes in color. Simply by telling the machines to randomly pick one of three colors, with probabilities supplied by the user, any color is possible. With global coordination, the machines could display any image the user desires, in full color, with micron resolution.

Force

Materials use force to change shape, in some cases to maintain shape, to absorb acceleration and deceleration, to vibrate, to move external objects, and to move themselves. Since often-expensive energy is needed to generate a force, we'll consider three situations: those requiring little to no energy, moderate amounts, and large amounts of energy.

Applying a force always requires energy from some source. So when we say we need little or no energy, we really mean we're being efficient about recycling energy that came in from outside, or finding ways to replenish the energy supply over time. For example, nanomachines can act as springs to absorb and then return energy—but not necessarily immediately, as a normal spring does. This can lead to materials that deform when pressure is applied, but return to their original shape when the pressure is relieved, either immediately, over time, or on command.

With moderate energy, nanomaterials can significantly reshape themselves, change their texture or optics, or provide vibrational or other dynamic stimulation. The energy required might come from outside the material or from the user. For example, if you shake a "book" in a certain way, it could display different content.

If energy is abundant, almost anything is possible. The material can reshape itself at will, carrying with it the user, other objects, or, for that matter, the entire house.

☐ Summary

Try to imagine living before the telephone was invented, or before there were computers, or before highways, or before plumbing—or before all of these. We are in the same situation today with nanotechnology.

The point of this chapter is not that any one of the particular items described will change your life. They all will. While it is quite possible that none of these uses for nanotechnology will be desirable or even feasible, it is virtually certain that an equal number of uses, over an equally far-ranging set of applications, will be invented and developed. Perhaps more than any other technology invented up to now, nanotechnology is all in the details.

■ 6 The Companion: A Very Personal Computer

John Papiewski

The appeal of seeing society's data structures in cyberspace . . . is like the appeal of seeing the Los Angeles metropolis in the dark at 5,000 feet: a great warmth of powerful, incandescent blue and green embers with red stripes beckons the traveler to come down from the cool darkness.
—Michael Heim

The Street finds its own uses for things.
—William Gibson

☐ A New Dawn

Compact, powerful communications devices have been described in both science fiction and serious speculative technical writing. The twentieth century, with the advent of inexpensive, miniature electronics, has glimpsed possibilities of personal communications. The twenty-first century will dawn with the development of nanotechnology. Advances in communications technology will accelerate even more, and devices such as the Companion, described here, will become possible.

☐ Purpose and Scope

The Companion is designed to provide information, communications, and entertainment of unprecedented quality and quantity. It will include an extensive library, comprehensive two-way communications, high-quality audio and video displays, and powerful computer facilities.

Molecular-scale manufacturing will afford a tremendous advance in data storage capacity over current integrated circuit technology. It will be possible to create a pinhead-sized library with a capacity of several million trillion data bits. The library in the Companion will contain digitized versions of millions of books, thousands of musical recordings and motion pictures, interactive courseware on many subjects, and several special-purpose knowledge bases to help us in everyday life.

☐ Design Characteristics

- Personal empowerment, via communication, entertainment, information
- Lightweight, unobtrusive, attractive, durable, convenient
- High memory capacity, redundancy in communications and power supplies
- Several easy-to-use interfaces
- Realistic audiovisuals
- Trustworthiness through privacy and security
- Low cost and wide availability
- Credible as a future product; an outgrowth of currently available technology
- Easy to understand, not dangerous or threatening

☐ Appearance

Packaged as an ordinary looking pair of eyeglasses, the Companion uses the power of nanotechnology to deliver the services listed in the previous sections. The lenses of the glasses contain an imaging system to present a high-quality display. Integral earphones provide audio. Built of advanced materials, the Companion will be tough, durable, and lightweight. The Companion, like present-day glasses, will be available in a wide variety of styles and colors.

☐ **Technical Details**

The following sections detail the fundamental architecture of the Companion.

Frames

The frames of the Companion have ample room for data processing and power supplies. Data is stored and manipulated in a few cubic millimeters of nanoelectronics and nanomachines. Power supplies will occupy up to 90 percent of the available interior volume of the frames and more than 50 percent of the available exterior surface area.

The frames are likely to be a dark color to collect solar power effectively. The surfaces touching the skin have a slightly sticky material that gently clings to the face. This prevents the annoying sliding-down-the-nose problem—without invoking the parts-that-pinch problem. The inside surfaces have biosensors to pick up pulse, skin temperature, and blood chemistry via perspiration.

Audiovisuals

An imaging system embedded in the eyepieces produces extremely high quality pictures. The Companion uses phased array optics (PAO), described by Brian Wowk in chapter 9. Images produced using PAO can be indistinguishable from reality, given enough data for the picture. Conventional imaging would put the picture too close to your eyes to focus properly, but PAO can create an image that seems to float a meter or two ahead of you, allowing your eyes to focus on it naturally. The imaging system is mounted on a thin film inside the lenses. It rolls up like a window-blind to permit unobstructed vision and clear eye contact when special imaging isn't needed. PAO, when in place, would reduce incoming light about as much as a pair of sunglasses.

Not everyone will need corrective lenses to see clearly, and twenty-first-century medical science may have solved vision problems permanently. Depending on the owner's needs, the eyepieces can be either optically neutral or lenses.

Mounted right above the eyepieces are two video cameras. They are the same distance apart as your eyes to give recorded images a greater sense of realism. The cameras can be tuned to pick up infrared and ultraviolet as well as visible light. They can also be designed to provide light amplification for use in near-darkness.

Earpieces are provided for each ear, which resemble hearing aids embedded in eyeglass frames. The technology for the earpieces doesn't need to be much more sophisticated than in present earphones, though they will sound better and use less power. Miniature microphones placed near the earpieces pick up stereo sound from the outside environment. As with the video cameras, the microphones can be designed to perceive a greater range of frequencies and intensities than our ears can.

Computing

The Companion employs both neural networks and conventional logic systems for computing. Computer neural networks duplicate some functions of natural nervous systems and are well suited to learning and recognizing patterns (a face, a sound) and linking one piece of information to another. These don't need to be as powerful or as well organized as brains to be useful in a specific role.

In the Companion, neural networks analyze torrents of broadcast and sensory data and alert other components when something interesting comes along. Serial logic systems are like today's familiar computers. They complement the neural networks by being precise and reliable. They handle internal housekeeping, library, security, and other functions. The computers in the Companion are tiny: depending on requirements, hundreds to thousands of them may be employed.

Nanotechnology will enable a gigantic library to fit into a tiny space. Data storage schemes, such as coded polymer chains, may be able to achieve storage densities of a billion bytes per cubic micron, or one million trillion bytes per cubic millimeter.[1] This storage density is roughly comparable to DNA, which packs about 100 megabytes into a small part of a cell. The Companion's design assumes a one-cubic-millimeter block of such memory, however, if this isn't enough, the frames have room for several cubic millimeters.

This data storage capacity is astronomical by today's standards, exceeding the present combined storage capacity of every computer in the world. The problem with capacity advances, however, is that the marvelous gains have been temporary; people are quick to find new uses for expanded memory capacity and fill it quickly. The libraries and software described here are based on present-day trends; the needs and wants of twenty-first century society will doubtless make these projections seem naive. The components described will occupy less than a tenth of one percent of the Companion's memory, leaving the rest free for a lifetime of use.[2]

A self-integrity system consisting of dedicated computers and software keeps tabs on the Companion's battery condition, structural integrity, available memory capacity, and security status. It will automatically alert a network-based emergency center if it or its owner comes to harm. If it is lost or stolen, the personal data contained inside can be automatically erased or transferred to another Companion.

Communications

The Companion uses both optical and radio channels of communication. The outside surfaces of the frames are lined with dozens of optical send-receive units; a radio antenna 20 centimeters long is imbedded within the frames. The light-based link is by far the faster medium, capable of sending and receiving several billion bits per second, but it will be limited to wherever you can find an optical outlet. Radio is more widely accessible, but slower. Because of noise, power, and distance trade-offs, radio channels will have a much smaller bandwidth for transmitting, less than 500 kilobits per second. This will be sufficient for voice transmissions or small data files.

In the next ten to twenty years, fiber optic outlets will likely replace conventional telephone and cable TV in most homes and offices. If you're at home or in a public building, your Companion will have the full high-speed access that optical fiber offers. Optical outlets placed on the walls and ceiling could bathe most of the room's volume in low-power infrared light. These outlets would be on all the time. When you walk into a room, your Companion would sense the infrared signal and send its own signal, establishing a two-way link. The Companion's frames will have optical receivers all around, so turning your head or moving around the room won't break the connection.

If an optical outlet is not close by, radio channels are still available. Today, a portable phone provides service as long as we're close to civilization. With low-orbit communication satellites like Motorola Corporation's Iridium going up, cellular radio services will be available from anywhere on Earth.[3] You'll have to leave your Companion at home to be out of touch.

Today, a radio or television tunes only one channel or frequency at a time because it has only one tuner circuit, whereas home VCRs allow you to record one channel while you watch another. Molecular technology will make it possible to build hundreds or thousands of tuners in a tiny space, all working simultaneously. Each tuner would have its own computer that selects and records broadcasts the user might find interesting, and notifies the user when the recordings are ready. Using this technology, the Companion can extend the user's reach into information and entertainment channels tremendously.

Power

A variety of sources such as photovoltaic cells, fuel cells, rechargeable batteries, and thermal-differential cells provide power. A direct electrical connector will be provided for recharging if the batteries become completely drained. Several different power sources are necessary for two reasons: if one source—say, the solar batteries—isn't sufficient in a situation, such as darkness, other sources will take over. The other reason is that the total aggregate supply must be up to handling peak power drains on the system.

Power is the biggest limit to the Companion's capability. Radio transmission consumes the greatest amount of power, up to perhaps five watts. The display, earphones, and optical transmission each use less than 100 milliwatts for continuous duty. The computers and memory draw a negligible amount of power, less than one milliwatt. The Companion may be limited to a maximum of an hour or two of radio frequency broadcasting before automatic monitors prevent any further power drain. The power supply would have to deliver a maximum of about five watts, and the battery capacity needs to be at least 10 watt-hours.

Interface

The spoken word will be the cornerstone of the Companion's user interface. It will handle such casual requests as, "I want to call my sister Susan," or "Is there anything worth watching tonight?" Continuous speech recognition, a long-desired goal of computer scientists, has seen some dramatic breakthroughs in recent years. Personal computers will increasingly incorporate this technology in the next few years, and common appliances are likely to incorporate voice recognition soon after.

Since speech isn't always appropriate or convenient, the Companion provides other interfaces. Eye-motion tracking sensors in the frames combined with projected visual control panels allow the user to operate the Companion through eye movement.[4] This virtual display panel can also be operated by hand movement; built-in cameras locate the user's hands and fingers. When needed, realistic images of keyboards, levers, buttons, and knobs are projected into the user's field of view. The projected controls move and react the way real controls do. The controls do not provide force feedback, but an audible click or other sound effects make up for this shortcoming. Much of this technology is already available in the "heads-up" display helmets that fighter pilots use today, which eliminate the need to look down at a control panel while in flight.

☐ Library

The Companion contains an extensive library of information and entertainment, including digital versions of books, music, movies, and more.

Data Bases

Books can be represented in a computer as graphic images of pages or as text, or both. By using images, the original look of the book is preserved, including typefaces and artwork. But images take up a lot of memory and are difficult, if not impossible, to search. Plain text is memory-efficient and searchable but a little dull. By using both techniques, millions of books can be stored.

In addition to books, the Companion's memory has enough room for many thousands of hours of video. Since many thousands of still art works fit in the same amount of memory as an hour of continuous-motion video, a large still art collection will be offered as well.

Personal empowerment is a key goal in the design of the Companion. To that end, the library will be well stocked with interactive computer software called courseware, which combines various media (narration, music, text, video, quiz sections) to educate the user in subjects ranging from first aid to French cooking. An owner's manual for the Companion will also be included.

The companion will also be able to play many games. Old standbys—card games and chess—are offered along with twenty-first century video games and interactive electronic-community events. Foot pedals, joysticks, and other equipment can be interfaced easily. Today, thousands of computer hobbyists across the country link up nightly to play shared games on computer networks. Games like these also will be possible with the Companion, but with far greater realism.

The library contains extensive and detailed maps. The maps combine with the computing and communications facilities to form a powerful navigation system. The global positioning system (GPS), a present-day satellite-based navigation system originally developed for military applications, allows a radio receiver to determine its location, accurate to a couple meters, anywhere on Earth.[5] Thirty years from now GPS, or its eventual successor, will be able to locate any constructed object large enough to contain the required antenna—about ten centimeters. The Companion can receive these signals to determine the user's location as well as the location of other users and specified objects.

The personal library contains audiovisual recordings the user makes, both live and copied from broadcasts. It also includes a personal medical history and a record of individual preferences in entertainment and news. Directories, notebooks, and active calendars keep track of friends, relatives, acquaintances, business associates, important dates, and other changing information.

Knowledge Bases

The Companion contains knowledge bases for medicine, law, and finance. *Knowledge bases* are sets of facts, procedures, and rules of thumb that pertain to a given field. The medical knowledge base provides a multimedia resource covering anatomy, medicine, and psychology. It emphasizes self-help. And it ties in to the owner's private library to provide an ongoing personal medical history using data gathered from the biosensors. In case of an emergency, the Companion can automatically call for medical help.

The legal and financial knowledge bases provide the user with extensive advice on tackling business and life's sticky situations. Like the medical knowledge base, the financial and legal knowledge bases tie in to the private library that keeps track of the user's personal financial and legal data.

A survival knowledge base helps the owner cope with severe and life-threatening situations. It describes how to gather food, build shelter, treat injury, and deal with terrain and wildlife. Of course, with global communication capabilities, sending for help will be easy. Yet some people will want to challenge themselves by setting out deliberately with only the Companion's suggestions to show them how to survive. Wilderness training and survival courses enjoy popularity today because of the challenge and escape such activities provide. As technology progresses and life gets more complex, some people will crave the direct confrontation of physical survival more and more.

The library includes extensive language translation facilities and a cross-cultural etiquette knowledge base. The user can carry on a conversation in a foreign language and the Companion simultaneously translates a foreign tongue and presents it as displayed text or spoken words. Automatic or machine language translation is techni-cally very challenging, but even if imperfect, machine translation can still be helpful in many situations.[6]

The etiquette knowledge base contains information on cultural practices, protocols, and taboos. With this information, users can find out what's appropriate to bring to a birthday party in Warsaw, Ohio, or a wedding in Warsaw, Poland. Although you may never use the etiquette knowledge base in a practical situation, many people will find this information fascinating, and it will help us appreciate the way others live.

☐ **Additional Features**

Security is provided at many levels. An identity file containing a retinal scan, blood chemistry profile, and speech patterns personal-izes your Companion. This identity file is developed over time as you use your Companion and is continuously updated. It can be used to prevent another person from activating a stolen Companion. In addition, you will be able to give explicit commands and pass-words to make your personal information more private.

Computer viruses, "Trojan Horses," and other methods of hacking pose a greater problem than outright theft. Powerful computers in the larger communications system and in the Companion will prob-ably be dedicated to watching for suspicious content in received data. Since some people want to protect information, and others want to get at it, data security is likely to remain a very tough issue, regardless of the technological milieu.

One of the greatest advantages of having data computerized is that it can be searched rapidly. Techniques range from brute force meth-ods that find every mention of the poet Robert Frost, to artificial-intelligence searches that spot poetry in a Robert Frost style, and hypertext searches that let you know what Robert Frost's relatives did. Music and pictures can also be searched. At a minimum, de-scriptive text will accompany each movie and musical selection, which can be searched by commonly used text search methods. Neu-ral networks and advanced AI may be able to search the actual content of the selections for musical phrases, scenes, and the like. This would be like using the speed-search button on your VCR, if the VCR was smart enough to look for scenes you wanted to see.

With few exceptions, all this material is already available in computerized form. Maps, books, and music have all undergone computerization in recent years. Video and movies will follow as the cost of memory continues to plummet. The main contribution nanotechnology makes to the Companion is data storage big enough in bytes, yet tiny enough in volume, to carry all this material around with you, easily.

☐ Alternatives

The hardware technology and packaging, interesting though it is, is only secondary in importance to the libraries and working software of the Companion. The physical form of the Companion, eyeglasses, is a convenient package because it gives easy access to the eyes and ears, is lightweight, is big enough to contain the necessary hardware, and is already an everyday sight.

But the Companion's technology may be delivered in other ways. One possible configuration would be similar to a pocket calculator with a foldout viewing screen, similar to Apple's Newton. Or the Companion could be designed as an implant, inserted into the body, powered by blood sugar, and connected directly to visual and auditory nerves. Assuming it will be harder to work out these biological details than to use known technologies for visual displays, audio, and so forth, an implant version of this technology—while inevitable—seems further in the future.

One style alternative that may be quite attractive is wraparound goggles that would cover the entire visual field, producing an artificial-reality effect that is much more compelling.

☐ Marketing and Costs

One of the promises of molecular manufacturing is to drastically reduce the costs of making things. In the Companion's case, the costs of design will probably be the greatest: organizing the data, designing the hardware, and writing the software will take a substantial effort.

But the market is huge. Potentially everyone who has a phone or a TV set will want one. So the costs of design and manufacture can be amortized over hundreds of millions of units. This would make the Companion so cheap that it could be sold for a very nominal amount, say, one dollar. The real cost to the end user would be determined by the services he or she chose to subscribe to.[7]

☐ **Social Impact**

So, where does all this fancy technology get us?

Media

To begin with, radio, telephone, and television will merge into a single media resource. The familiar paper print media of newspapers and magazines will be largely displaced; print reporters and editors will gradually move to this new medium. Social forms of information and entertainment (cinema, night clubs, town meetings) may remain but probably in an altered form.[8]

For much of the history of broadcasting, the major networks were the only game in town; people had few choices in what they could see or hear. Recently, cable services and low-cost videotapes have made huge inroads in the networks' formerly sovereign territory. This trend toward entertainment on demand will most likely continue. Individual tastes will become more specialized, more esoteric. The Companion is well suited to cater to specialized tastes. Its huge storehouse of entertainment and its high-speed communication system allows entertainment consumers an enormous selection from which to choose. In addition, powerful search capabilities will be able to sort through a large selection quickly and automatically.[9]

Politics

Because television broadcasting is still controlled by a few well-heeled interests, every society, from the most liberal to the most reactionary, has this medium manipulated to some degree by the government. The Companion puts the technology of television broadcasting in everyone's hands. It may no longer be possible for large institutions to deliver an impenetrable wall of distorted information. If eyewitnesses are present, "facts" will be extremely difficult to control. Currently, personal communications, from letters to videotapes, are seen as politically subversive in some areas of the world. In such areas, devices as powerful as the Companion would be officially banned, though enforcing such a ban would be extremely difficult due to its cheapness, small size, and innocuous appearance.

If social acceptance is high, the Companion has the potential for becoming a new political tool. The Companion's software could include an electronic voting booth. Since the Companion will know the user's home address, it will present only those candidates and issues appropriate for his or her political district. Information and propaganda on the pros and cons of issues can be called up for display. The Companion sends its electronic ballots to a central voting system. Voting data could be encrypted to preserve vote secrecy. Vote counting would be very fast. Today, informal radio and TV polls are taken by people calling in and registering their opinions by Touch-Tone phone. These public opinion polls, while entertaining, carry little political weight. The use of the Companion in elections and referenda could change the face of politics forever.

Everyday Life

It's tempting to think that, as radio squelched the art of conversation, and television has assumed a central role in many of our lives, the Companion also will be divisive in personal relationships. Since it provides so much information at a personal level, it will undoubtedly help people become more independent. By catering to the most esoteric personal tastes, the Companion will make our already fragmented society even more so. However, although it's true that many people may become even more addicted to a never-ending stream of news and entertainment, remember that the Companion is a *two-way* communications device. It's designed to connect like-minded individuals and to aid social interaction with its etiquette and language libraries.[10]

People will use their Companions to record their life's experiences as they do today with still and video cameras. It has such a voluminous memory that a person can, for example, let it record an entire two-week vacation, then review the results later and (we can only hope) weed out the boring parts. One possible downside is that with all these cameras on all the time, people will be more circumspect about how they act in public, lest they be caught in some terribly embarrassing moment. "Candid Camera" and "Funniest Home Videos" may be just the tip of the iceberg. For this reason, it may become a social gaffe to wear your Companion to any gathering where mutual trust and privacy are important.

Because the Companion contains a highly personalized data bank that is tied in to communications networks, the potential exists for institutional abuse. Governments could use the Companion to keep track of political undesirables or foreign nationals. If the security were broken, a companion could function as a portable spy even without the owner's knowledge. Demographics brokers could secretly gather buying patterns and personal preferences to build targeted advertising lists. These unfortunate prospects of the device will lessen its acceptance by some. The Companion will have security features, but even so, many people will be reluctant to own one out of fear of having their privacy invaded.

All too often, advances in technology promise to make life easier but end up delivering only fleeting glamor and then endless complication. The Companion will be exciting, complicated, and will make life easier. Serving as a personal secretary, it will remind us of appointments and keep track of all those phone numbers we need to know. For the most part it will be unobtrusive, doing its own internal housekeeping, requiring no battery replacements, and bothering us only when we ask it to. The real challenge is deciding how to use such enormous informational resources. Undoubtedly, our values will shape how we use it, and using it will in turn shape our values.

☐ Handicap Assistance

The wonderful possibilities of advanced molecular and biological technology give us the hope that we might eliminate physical handicaps. However, this will likely be very difficult to achieve; many of us may be blind and deaf for some time to come. Nevertheless, the Companion, while not warm and fuzzy like a seeing-eye dog, can be of service to the handicapped.

The blind can benefit in several ways. Neural network and pattern recognition technology may be sufficiently advanced to program everyday objects into a visual memory and to allow an ongoing learning process for unfamiliar objects. The Companion's video cameras, surveying the scene, can provide a running verbal narrative about the scene before you (like the Simultaneous Audio Program channel on contemporary TVs).[11]

Likewise, a hearing-impaired person could "listen" to someone talking. The Companion would pick up the audio and translate it in real-time into a text or sign-language window displayed in his visual field, like closed-captioned television today. Discriminating the right sounds may be a problem though: picking out one voice in a crowded or noisy room may be as difficult for the Companion as it is for a person.

People with a physical handicap that limits their movement could use the Companion as a control center. Devices such as appliances, doors, robot arms, and the like can be designed to be controlled by the Companion. Using eye movement to drive these interfaced devices, a person could run a household of automated equipment even if he or she were completely immobilized.

If biotechnology doesn't solve the problems of aging, the Companion can also help the elderly. They can either summon help at a moment's notice or have the Companion summon help automatically, sending a signal if the user's vital signs look bad. The Companion could also aid their hearing and assist visually, if needed.

Normally healthy people can have their senses enhanced as well. The cameras and microphones both cover a broader range of light and sound than our natural senses can detect. The Companion can be used with infrared or low-light capabilities to find your way around in the dark, or with audio sensitivity boosted to listen to faint sounds.

In combination with the appropriate software, the sensory apparatus can do much more. The microphones can detect a specific sound from a jumble of noise. The cameras can pick out an object and zoom in on it. This would be an obvious boon to bird-watchers, since the bird's call can be automatically isolated and identified. In addition, it could help us all out of those terribly awkward moments when someone recognizes us and calls us by name, but we've forgotten the other's name!

☐ Barriers and Breakthroughs

Technical

The Companion depends heavily on nanotechnology. With the present level of technical activity, it seems likely that nanotechnology breakthroughs such as assembler construction may take place as early as the year 2010. The Companion's display, communications, power supplies, and computers are all refinements of currently available technologies. The only effort would be to translate these known technologies to a new form, similar to designing a radio using transistors instead of vacuum tubes.

The Companion assumes only moderate advances in current communications technology; it requires no major breakthroughs. Multi-channel radio frequency transmission has been in practice for the better part of a century. Fiber optic communications development is well underway and will continue to replace copper wire as the bulk communications medium of choice; we can expect to see fiber optic connections in some homes by the turn of the century.

Social

With television and radio slowly migrating to cable and other media, slices of the airwaves, or radio-frequency spectrum, will be freed up for other uses. The process of deallocating and reallocating the airwaves isn't a free-for-all; it's heavily government regulated and requires years to hammer out each detail. The Companion would have go through approval by the FCC and the analogous agencies of other governments before necessary radio bands can be allocated.

Currently, the most difficult barrier to producing the Companion seems to be our unresolved struggle over intellectual property rights. Digitized books, movies, and music must be created somewhere. Much of this material is currently in the public domain, and more will be by the time the Companion is developed. But under existing copyright law, newly created works will be protected for as long as seventy years. Copyright holders will naturally want compensation. The legal and economic process of putting all this information and entertainment into the public's hands will be quite complicated.

If we are unable to establish a technical means for rewarding the holders of existing copyrights, the Companion can still be sold with a standard library of public domain works, and the user would have the option of purchasing copyrighted works from various vendors. The user, of course, won't have to decide at the time of manufacture; data packages can be transferred into the Companion easily at any time over the radio or optical channels.

Pay-per-use is another alternative. A huge library could be delivered including all the desirable movies, music, and games, but some material could be locked out. You would call an entertainment service that unlocks the requested items and bills your account. This is now done with pay-per-view television and software delivered on optical disk.

In addition to copyright complications, legal and medical professionals, fearing an impact on their careers, may resist providing cheap availability of knowledge bases gathered from their fields. This fear may be unfounded; many self-help books on medicine and law already provide help for individuals without adversely affecting these professions.

☐ 2020 Vision

The most formidable breakthrough needed to produce the Companion will be the development of nanotechnology itself. Computerized libraries of information and entertainment will emerge with or without nanotechnology, and the process is already well underway. Based on these trends, we can project the Companion to arrive on the scene around the year 2020.

■ 7 Trivial (Uses of) Nanotechnology

H. Keith Henson

What mighty contests rise from trivial things!
—Alexander Pope

☐ Trivial?

Nanotechnology changes just about everything—or does it? Substantial numbers of Amish and Mennonites living in the United States stick as close as they can to a way of life frozen over a hundred years ago. Advancing technology has changed little for these folks—they still buy buggy whips! In the coming nanotech era, some individuals or groups of people might try to live a life based on current standards, using a minimal amount of the new technology. Advanced nanotechnology will give us the tools to alter the molecular make up of our own bodies, thereby generating a range of variously viable species, but if some of us choose to remain biologically human, it is fairly easy to imagine goods and services that we would want, and how "trivial" applications of nanotechnology might provide them.

Trivial applications of complex new technologies can result in a substantial business. People who purchased the first computers would, at the time, have been aghast at the concept of children using many millions of computer instructions per second (MIPS) to play silly games. Last year Nintendo and Sega split a "trivial" five billion dollar market! The same is true of word processor programs running on today's average (tens-of-MIPS) PC. By the standards of two decades ago, a powerful computing machine is being used for the ancient and relatively simply task of written human communication. Those who first ran computers with such capability would have been absolutely appalled at the prospect of a machine being used in such a trivial way, spending almost all of its time waiting for a keystroke. A simple mechanical typewriter would suffice for the task. Even a pencil would do. Using a million-instructions-per-second machine to type is technological overkill—but cost effective.

☐ **Gasoline Trees and Roving Real Estate**

If we consider future applications for nanotechnology, a similar overkill use would be a "gasoline tree," or, at first, an organic waste-to-gasoline converter. (I presume cars will still be driven for fun if nothing else.) In the early stages, such a device might be the size and shape of a washing machine. You lift the lid and drop in food waste, paper, cardboard, chunks of wood, and so on. An ash hopper off to the side fills up with ceramic marbles, excess water runs down the drain, and carbon dioxide comes out a stack.

Organic materials placed in the machine would be rearranged by nanomachines forming liquid fuels suitable for use in internal combustion engines. Actually, everything in this idea could be achieved today on an industrial scale; the chemistry and physics are sound, it just needs a little engineering development to put it into the home.

Later versions might come as a seed. Plant it next to the driveway, and it grows into a nice looking tree, complete with a recessed filler hose. Instead of continuing to grow more tree after it matures, it makes gasoline and stores it in the trunk. In fact, a vine that grows in the Amazon produces enough oil that oil can be skimmed off the tapped sap and used to run diesel engines. So, a gasoline tree might be possible to develop simply by using genetic engineering. And if you wonder how roads would be maintained without fuel taxes being paid, consider a self-repairing road that used solar energy to grow more road in place—kind of like crabgrass. If the "road plant" were partly derived from kudzu, road crews might be less busy with repairs and more concerned with keeping the road plant from overgrowing everything!

Providing energy is another trivial use of nanotechnology—or at least, providing the levels of energy currently used in the West. It is a good bet that solar collectors that are close to 100 percent efficient can be made with nanotechnology (well, ok, so they only hit 50 percent). Moreover, they could be self-repairing, grow from a seed like the gasoline tree, and double as permanent roofing. To install the solar collector, you would place a small square on your roof. A week or so later, it would grow an extension cord into your electric box and start supplying part of your electricity—the inverter is part of the distributed design. More elaborate models could supply a steady supply of electricity by growing an energy-storage module underground. Dense housing (if people still live that way) and large-scale industrial activities (whatever *they* are) will still require more energy than can be supplied by local sunlight, so good reasons remain for keeping the power grid operational. (Personally, I enjoy seeing transmission towers, but if people want to bury them underground, that can be done easily. Transmission lines can be made superconducting as well.)

Regarding transportation, it is not entirely obvious that the conventional needs of a community for getting around—such as commuting to work—make sense in an era of nanotechnology, but if people want to congregate in large numbers (for rock concerts?), it would be a simple matter to provide a personalized subway stop for every house that wanted one. With a little notice, a new line could be run wherever and whenever a large group of people needed to be transported to a new location. A hotel/convention center could consist of a big tank of undifferentiated "nanostuff," a large area of unoccupied land, and a subway terminal. Choose one of thousands (millions?) of designs for your convention.

Another possibility is movable real estate. With hordes of minute earth movers, surface structures could be moved slowly to open up new space for public works (again, whatever that might be in such an era). If an evolving community wanted to live in a less dense mode, all the structures, trees, roads, utilities, and the like could creep outward at an inch or so a day. (Talk about urban sprawl!)

A local version of the same idea would be timesharing part of your local real estate. A week before your big party, your house and lot would start to grow, while the neighbors' houses would shrink and move away. By the time your party began, your house would have metamorphosed into whatever size mansion was needed. Over the next week, it would shrink to its original size—or perhaps a little smaller, if someone else were planning a party.

☐ Home Appliances

Inside the home, there are countless applications for trivial nanotechnology. To start, the whole house could be carpeted with a self-cleaning rug. An early version would ripple like the cilia in your lungs to move all the little stuff that fell on it to a central bucket. Coins, nuts and bolts, paper clips, and the like could be sorted from dirt, dust, hair, skin flakes, and similar debris. A later version might be smart enough to know what it should eat. (It wouldn't do to have the rug eat the fur off one side of the cat while she took a nap!) A still later version might deal with the classic clothes-on-the-floor problem that has broken up so many marriages. Left for more than a few minutes, underwear, socks, even dress clothes could be disassembled by the carpet and rebuilt—clean, of course, and folded—in dresser drawers, or hanging in the closet.

Clean floors are even more of a problem in the kitchen. Carpets for kitchens and bathrooms (and in all rooms for those with pets or babies) could also have soak-up powers. A network of pumped veins within the carpet could either be connected to the drain, or (depending on how you felt about it) attached to the food synthesizers. A nanotech carpet could be covered with cilia so short that it would look and feel like a no-wax surface. An active no-slip option could hang on to your shoes and keep them from slipping while offering no resistance to vertical movement. (The same technique could be used between tires and the roads to get greater-than-one-g performance from vehicles.)

Nanotechnology would also let you redecorate—daily. Walls and ceilings would get nanotech paint that could act as a universal display screen programmed to appear however you wanted, up to a very high definition TV screen. Talk about trivial, you could have striped paint! Don't like the color, or the art work? No problem: any color, any design (delivered by optical fiber) can appear at your whim with a few clicks on whatever is being used as a computer interface. (Naturally, there will be a small charge for displaying copyrighted art works.) You could even have a real-time view of the Grand Canyon on the wall of your living room.

Furniture could be ordered in the same way. Static items that you wanted to keep for a while could be built by your general-purpose household constructor. More dynamic items—an end table today and hassock tomorrow—would be made of general purpose nano-stuff, sort of an "Erector Set-in-a-bucket." And don't be too surprised when you come home to find that the kids have reprogrammed the spare bed into a slobbering, eight-foot-high Godzilla.

Structurally, houses could be reinforced to the point that no natural disaster—short of a large incoming meteor—would do much damage. Certainly earthquakes, tornados, hurricanes, and the like could be reduced to minor annoyances. When buildings are reinforced with intelligent diamond fiber, such events would cause little or no real damage. And if damage were done, a nanomachine-based house could repair itself.

If we choose to retain historical structures "as is," we can coat them with diamond, and all deterioration—termite damage included—would be stopped. Actually, every prenanotech structure might be considered historical, but most of them will be replaced, or greatly modified. If there is any question about the historical value of a structure, we can make a record of how it was put together (down to the marks on the nails). Then we can always reassemble it later.

With all the new comforts of home, will people leave home very often? This is a hard one to answer. Even now, a growing number of people choose to work at home and telecommute. In the future, if you "work" at home (whatever that is), and if tasty, nutritious food is made in a "cabinet beast"—from elements scrounged off the floor or out of the air, and rearranged with energy from your rooftop—and if most of your shopping needs are met with home-shopping networks that deliver instruction sets by optical fiber to the nanomachines in your house, most of the reasons for going out vanish. Perhaps this is the real solution to the traffic problem!

☐ Health Assurance

With all the time on our hands from less commuting, would we spend more of it worrying about our health? I doubt it. One of the primary reasons to develop nanotechnology is to cope with nasty health problems—including aging. Once we have done that, those who stick to a more-or-less standard human configuration will want improvements. Self-repairing, diamond-reinforced teeth for example. Or replacements for glasses. We are already reshaping eye lenses with crude surgery. Actively controlling lens focus would be an obvious application for smart nanoimplants. Floaters could be cleaned out of the eye fluid by nanomachines designed for the purpose.

A more complex modification would be rearranging the retinas of our eyes. Why evolution wound up with vertebrate nerve circuits in front of the light receptors is not well understood.[1] We get along fine with the wires on the "wrong" side, but what a trivial use of vast developments in biological nanoscience to be able to turn the retina over in place. In that way we could get rid of "unsightly" blind spots. A more challenging problem is that of giving sight to people blind from birth. Building eyes may be easy compared to rewiring brains.

Given an understanding of biology at the nanolevel, nanotechnology should be up to solving the subset of human health problems due to mechanical or chemical causes. The same will be true for our pets. Given the rate at which new treatments are approved, our pets may get it first. I expect there will be a demand for animals that can be switched on and off. ("Honey, did you remember to turn off the dog?") The biggest difference for humans between horses and cars is that cars don't need attention every day. Members of the Society for Creative Anachronism (SCA) really need switchable animals. Their battles and pageants are short on horses, but few of them want to take care of a horse between events.

In addition, the people get so banged up in mock battles that the SPCA would object if real animals were subjected to the same beatings. Once a thorough understanding of human anatomy and chemistry is stored in nanomachines, the SCA can stage real—dripping-with-gore—Conan-style battles. After the performance, the chunks of the participants can be stuck back together, Valhalla style. Not my idea of a good time, but I don't much care for football either.

While there seems to be no limit to the level of Medieval realism that could be achieved, we don't have to put up with vast amounts of dung. A partial solution to the dung problem has been proposed in the form of a "doggy afterburner." Such a device (critter? relative of the tapeworm?) would inhabit the lower intestine of an animal, burning the organics out of whatever came along. The nonorganic elements would be formed into ceramic marbles, to be excreted at rare intervals.

In a nanotech world, human relations with engineered animals could get very weird indeed. Today, pastoral nomads in Africa drink the blood of their cattle. A less messy method would be to grow plugs on the animals that could be connected to humans directly, supplying energy and materials to the bloodstream. Instead of killing sheep, you could bring in a batch and "recharge" from them. A lower-on-the-food-chain alternative could be a grafted on "back-pack" that would unfold when you lay in the sun, forming a large photosynthetic area. Assuming the nomad's sheep don't trample you, with thirty square meters of surface, a few hours a day soaking up rays would eliminate the need to eat animals or plants. Totally recycling "hypervegetarians" would be able to live the simple life par excellence—really leaving "nothing but footprints."

☐ **What Work?**

In a world of nanotechnological systems, what would people do for work anyway? Work is an activity specific to humans that got started long, long ago when we started moving food from where we found it to where we could eat it. Work has become so elaborate, the origins have almost been lost and have been made very indirect by the invention of money.

One of the difficulties we face with nanotechnology is the question, "What do you do with the money, or its equivalent, that you receive in exchange for work?" There are still many things you might want to buy—land, transportation, elements, energy, information, even finished goods. Is there reason to expect that any of these will be very expensive? I don't think so. There is a whole lot of land, especially with nanotech to make it worth something. It also seems likely that we could make a lot of really nice new land, reorganizing what we have here to make more beaches, or we could go into space. The reorganization can incorporate as many vacuum tunnels for transportation as we wanted. The useful elements are common and, with solar energy on tap, power for personal use can't be expensive. (Energy required to launch interstellar crafts may be another matter.) The kicker in a nanotech world is that no matter how much something costs to make in the first place, it costs almost nothing to duplicate, because it's mostly made of information. Finished goods will become "services" rather than distinct products. On the other hand, demand for personal services may actually increase, if you seek interaction with a real human.

We may see a long period of deflation offset, perhaps, by a reduction in the need for money to live. Or, at worst, you might get a photosynthesizing modification (see above) and live off the land. But what of money itself? Perhaps it will need to be based on chips that actively resist duplication, or possibly coins of genuinely rare elements. Or perhaps, continuing a growing trend, money will be no more than abstract patterns exchanged by computers.

Even if we don't need much money, what kind of work might be in demand? Is there a trend in the past few centuries (in this century?) that might continue? At the beginning of this century, well over half of American heads of families were farmers. Today, some two percent are so occupied. Are the rest unemployed? Far from it. They work in a vast array of jobs, many of them information related, few of which existed at the turn of the century.

If you lump communications, entertainment, news and journalism, scientific endeavors, engineering, and most of the functions of government together and call it the "information industry," it is clear that this segment of the economy has accounted for an ever increasing fraction of what people do for employment. It may turn out that the nanotech revolution hardly causes a hiccough in this curve. If your "work" creates new knowledge, stories, or entertainment of any kind, you are likely to have a job in a nanotech world— if you want one.

Government, too, faces uncertain prospects. My libertarian bent prefers little or no government, but my guess is that there will be a substantial fraction of the population working for the government, or perhaps living on welfare, provided with food, shelter, and endless reruns. Another trend in the service sector that gives me the shakes is more lawyers: the richer people get, the more legal actions they seem to be involved in.

☐ Real Wealth

Which brings me to my final point. There is no reason a nanotech world should not be very rich. Again, this is in line with long-term trends. Citizens of the United States who live in poverty today are (on average) better off in many ways (square feet of living area, nutrient content of food) than the average worker in the 1930s or those living under current average conditions in Eastern Europe. Food and floor space should be nearly trivial cost items in a nanotech world— at least for populations within an order of magnitude of our current population.

In an era of nanotechnology, it might turn out that people work because they want something to do, or are seeking status, rather than simply facing the harsh realities that say we must work or starve. More and more people have been deciding what they will do with their life by criteria other than just making enough money to survive. This may represent a long-term trend.

As we become wealthier, will we travel more? What if the entire solar system were open for exploration? My personal pet project is interstellar exploration—and parties. At a recent Asilomar Conference, I asked a room full of cryonicists how many would go on a several-hundred-thousand-year expedition to the far side of the galaxy. About 95 percent of them were ready to pack up and go.

Galactic dispersal of the nanotechnologically advantaged presents an interesting challenge that I will leave you with: How can we carry on economic exchange if the only thing exchanged is information carried on laser beams? Establishing a value for something where a bargaining cycle takes millennia is an intriguing problem. We may choose to send short-duration personality constructs designed to handle the bargaining along with encrypted goods. But would that be satisfying? And if so, for whom?

■ 8 Nanotech Hobbies

Tom McKendree

No man is really happy or safe without a hobby, and it makes pre-cious little difference what the outside interest may be—botany, beetles or butterflies; roses, tulips or irises; fishing, mountaineering or antiquities—anything will do so long as he straddles a hobby and rides it hard.
—Sir William Osler

☐ An Abundance of Hobbies

Western society has long been a culture of abundance, its people spending a great deal of time, money, and effort in "the pursuit of happiness." People go dancing, see movies, and visit amusement parks. They build model railroads, collect stamps, garden, and pur-sue crafts. Others explore more esoteric hobbies, like war-gaming, bungee jumping, and dressing as medieval lords and ladies.

Nanotechnology will affect hobbies as dramatically as it affects everything else. Far and away the biggest impact, however, will be from the unprecedented abundance that nanotechnology makes pos-sible, which will remove the need, for most people, to spend much time earning a living. People will have much more time to pursue hobbies—most of their lives if they so desire. Nanotechnology will also make possible more extravagant, gigantic, and imposing hob-bies. Here are some examples of the possibilities.

☐ Model Railroads

Nanotechnology is the most extreme form of miniaturization using ordinary matter, and because miniaturization is at the heart of model railroading, one might expect the two to be a natural match. A rotary engine can be at least as small as 30 nm in diameter, as demonstrated by bacteria that use such engines to move.[1] Thus, one could build a model railroad where the rails are 50 nm apart (a scale ratio of over a billion to one.) Because a single engine or car would be near a light wave in length, a train would be much too small to see directly. On the other hand, one could build a working replica—small but visible—of the entire U.S. railroad network.

Actually, model railroading is usually not concerned so much with making things small as with making them realistic at the selected scale. At any current model railroad scale, nanotechnology could realistically render the shading on a splinter inside the letter "e" carved into the back of a wooden seat in a passenger car. Every detail that could be visible in a model railroad could be created using nanotechnological tools, although with the danger that tiny pieces—such as HO-scale playing cards—might blow away.

Most model railroads include human figurines. Because nanotechnology will be able to produce motors down to a millionth of an inch across, and hundred-million MIPS computers that will fit in a Z-scale skull, figurines that move and act realistically could be built for every model railroad scale. Some model railroaders may be reluctant to have nanotechnology provide everything. After all, a great deal of the satisfaction in making layouts comes from doing the work oneself. There are two responses to such reactions. The first is that much of that work requires delicate tools, and nanotechnology can provide the tools to create finer details oneself. The second response is that some people who might want to start model railroading find the initial set-up too daunting; making the first steps easier might entice more people into the hobby.

Philately

Many hobbies involve collecting things, and stamps and coins are classic items to collect. How will nanotechnology be incorporated into stamps, and how will this effect stamp collectors? One might reasonably argue that with massive computer networks linking everybody, postal mailing will disappear. That is possible, but often technology displaces old forms without completely eliminating them. Western Union, for example, still sends telegrams. Assuming postal mail continues to exist, how might nanotechnology affect the stamps? There are many possibilities. A working model railroad of the entire U.S. rail network, mentioned above, could fit on a postage stamp. It would certainly be an original commemorative issue. Indeed, working models and detailed scaled miniatures of anything could be produced cheaply for inclusion in stamps. Stamps themselves might change so that each included a nanocomputer, with unique serial numbers to reduce copying. How valuable would a Lincoln stamp with the serial number 05.15.1865 (the date Lincoln died) be?

☐ **Numismatics**

Originally, coins had the value of the metal they contained. A coin was simply a way of certifying that a particular hunk of metal held the proper weight of gold or silver. Through many intermediate steps, we have come to have fiat money, which governments say have value, and which does have value when others accept the money in exchange. Thus, a fifty-dollar bill does not carry fifty dollars worth of ink and paper, but is rather a symbol for the wealth of fifty dollars. Nanotechnology will make possible atomically perfect counterfeits, so using money that is merely a physical symbol will no longer be effective. Instead, every coin could be given a unique serial number—but then a counterfeiter could pass different copies at different stores, unless a central data registry recorded who held what coins and bills. In that case, people might as well use the data registry and forego the actual coins and bills.

If this happens, the supply of new coins will end, but the demand to hold collected coins will likely stay at least the same, and may well grow dramatically when people have more time and wealth for hobbies. A possible answer to the counterfeiting problem is to return to coins that have value based on their content. The only way to perfectly counterfeit a one-ounce gold coin is to use an ounce of gold. This will be popular with some people, since not everyone wants the details of their finances kept in a central computer. If this sort of currency became predominant, it would turn the fiat money of today into interesting curios.

In the case of both stamps and coins, authentic pre- nanotechnological specimens will be rare relative to the wealth that follows, and may be prized as tokens embodying those older times.

☐ Copying Issues

Nanotechnology will allow one to make perfect duplicates with atomic precision. (To make convincing copies, forgers must not only put the same atoms in the same places, they must use the same isotopes as well. Otherwise, a forgery could be recognized if it had, for example, the wrong amount of carbon 14.) How will collectors know they have an original? The easiest way to distinguish originals from copies is to put a mark on the original. Today, invisible serial numbers are etched into diamonds. This approach will foil those who create copies from detailed photographs of an original, but it will not work if the forger can use the original itself in making a copy. One approach would be to create a registry to authenticate that an original is an original. For collectibles that predated nanotechnology, this would be persuasive—if you could verify that the authentication record predated nanotechnology. Anyone with a valuable collection today will likely want to prepare a complete inventory and, if possible, register that inventory.

On the other hand, not everyone can have an original. By creating reproductions of significant items, more people will be able to become collectors. For example, an indistinguishable copy would be good enough for many who collect comic books. (A commercially legitimate reproduction would likely have "copy" unobtrusively written all over it, perhaps in lettering eight atoms tall, rather than be truly indistinguishable, but this will be unnoticeable for anything besides checking authenticity.) A final difficulty—if one were to take an authentic Roman coin, disassemble it, and then reassemble two atomically precise copies, each with 50 percent of the atoms from the original, all in their respective original places, could one really say that neither coin was the original?

☐ Gardening

There are those who like to garden, and others who think it is just an outdoor chore, but that's what makes a hobby a hobby. One of the biggest problems for those in the first group seems to be pests. Every plant is a living thing, and there are always other living things that would like to eat it. Although it may be an incredibly complex task, it may be possible to build ecosystem protectors. Such devices would not need to replicate, but they could be designed to kill specific weeds or animals as effectively as any predator. Ecosystem protectors could guard against invading species such as kudzu, fire ants, and killer bees in the United States, or rabbits in Australia. The ecosystem a gardener is trying to protect is his or her garden. Nanotechnology could provide "garden protectors," to effectively kill off the unwanted pests invading one's tiny plot, without using chemicals or harming anything besides the pests. Or biological nanotechnology could be used to create new, hardy strains of plants that look and smell exactly like the original but are designed to thrive in utterly different climates. Thus, one could grow what looked like orchids next to a mountain cabin, and what looked like healthy blue fir on a sunny beach. The coloration of flowers, when they bloom, the fragrances, or any other feature of a plant could also be specified. One could even order personalized cuttings, so that each petal of a rose bore the cameo of someone in the family.

Any plant designed to delight or amuse must carry an extra burden that would put it at a competitive disadvantage in trying to survive in the wild. For added safety, however, these garden plants probably would not have any of the capabilities necessary to reproduce (although they could present asexual flowers.) There is one plant a gardener could grow that seems almost mythic, a wish-for-it tree. Once the tree had grown large enough, one could walk up and ask for any object—a cup, a bicycle, a computer, a bucket of utility fog, Companion glasses, or any other small object that could be made of available atoms and that the tree knew how to build or could design. Then, a bud would form on one of the branches. It would grow and grow, and eventually one could pick the desired object off the tree.

☐ Home Crafts

In a nanotechnological culture, home crafts will not be just an inexpensive way to build a coffee table or weave a nice table linen, but a means for people to express themselves in a tangible object and give something of their labor and human spirit to one another. Every object a person makes will have sentimental value.

Many people who like to build things have workshops in the garage because it is the easiest and least disruptive room in a house to convert into a shop. In a nanotechnological home, however, one could easily transform any other room into a workshop. On demand and with a little notice, the tools and work pieces could be put away, and the shop would turn into the old room again. The shop could have the finest hand tools possible with the sharpest cutting edge atoms can form. The handles could change their shape to fit the hands perfectly and to massage away muscle pains. Molecular intelligence on the work end would make smart tools that adjusted to the work piece, or that modified it at the atomic level, like varnishes, but with vastly finer control. One could also incorporate nanotechnology in the work piece, including, say, a video game in an ornately carved wooden board, or making a stool that was self-repairing.

Needlepoint seems to be immune to changing times. It is obviously not a cost-effective way to decorate cloth, so it is clearly serving other needs and purposes. One purpose is that needlepoint is much more personalizable than manufactured cloth; most gift samplers include spaces for names and dates for births or weddings.

In a culture of abundance, however, we will be able to provide cloth personalized to each individual thread no more expensively than bulk cloth is produced now.

The second function of needlepoint is the sentimental value mentioned above, and the final benefit is the pleasant, calming effect it has on those who do it. For this hobby, nanotechnology could help by removing some of the petty annoyances, by offering self-threading needles for example. A piece of cloth could have a pattern printed directly on it that faded away at each stitch, or could sound a gentle alarm if one mis-stitched a thread. For those who want to experiment with the final product, there would be active threads that move on their own, creating interesting visual effects in the final design.

Nanotechnology will make it incredibly easy to build specified objects, so designing furniture, pottery, textiles, and so forth might become a popular pastime practiced by those with artistic inclination. Final products being fabricated automatically.

☐ Thrill Seeking and Blood Sports

Not everyone spends their hobby hours in quiet seclusion; the adrenalin rush is widely popular. Witness bungee jumping. Molecular engineering could provide super bungee cords of varying thickness and much stronger material that would allow significantly higher jumps. Harry Chesley hints at (chapter 5), and Keith Henson describes (chapter 7), how to protect against sudden deceleration. Other approaches could work, and all would protect people who decided to jump out of airplanes without a parachute. Most daredevil activities are similar in that they have in common experiences of speed and acceleration. Future wealth will provide the opportunity for a great deal of both, and nanotechnology will provide the means to help people survive the experience.

Henson has suggested that hand-to-hand combat, using medieval weapons, could provide a thrill for those seeking historical entertainment in the future. This activity could be quite realistic, down to producing wounds that today would kill a person. One difficulty is that major damage to the brain may not be repairable even with nanotechnology. Those portions of the mind where information has been lost might not return on reconstruction. Truly severe head wounds could be fatal. Such games may well draw the more extreme thrill seekers, for whom nanotechnology may seem to have taken all the "fun" from life (defined by them as a real chance to die).

This brings up a moral issue: Should people engage in blood sports when their direct threat to life becomes small? This is not the question, "Should blood sports remain illegal?" With nanotech healers standing nearby, blood sports could be made less dangerous than sports car racing is today. If "play" is a result of the informed choices of consenting adults, many would consider outlawing that choice overly authoritarian, an infringement of rights. The moral issue arises when we consider that repeated exposure to brutality may lead to an acceptance of violent acts performed without mutual consent. If, as it seems, future technology will allow us to much more easily translate our thoughts into reality, we may want to cultivate finer thoughts than blood sports would lead us to contemplate.

☐ Thrill Rides

The classic thrill ride is the roller coaster, still growing in popularity. Higher performance bulk materials will make possible taller, faster roller coasters with more elaborate track layouts. With current materials, however, it is already possible to make roller coasters that pull "snap turns" sharp enough to kill people—current acceleration limits are based on rider safety, not the technological limits of roller coaster engineering. Thus, future roller coaster rides may not pull on people much more strongly than today. On the other hand, greater disposable wealth will mean that parks can afford much longer rides. Amusement parks have many other rides that spin, loop, lift, and twirl. A towering Ferris wheel made primarily out of diamond would look spectacular on the fairway at sunset.

☐ **Amusement Park Attractions**

Many people agree that the Disney Corporation offers the best amusement parks in the world. The Disney people insist, however, that the parks do not offer "rides," they offer "attractions." Anyone who has been to a Disney park can sense what they mean by the difference. A Disney attraction aims to be a total experience. The parks are divided into tourist-friendly representations of places one might want to visit: a land of fairy tales, the American Old West, or the future. Each attraction has a theme that fits within its area, and each attraction looks like it belongs there. The line one waits in for an attraction continues the theme while offering diversion. The attraction itself is entertaining and seems more realistic than just a roller coaster. Increasingly, other amusement park developers are trying to copy Disney's approach.

With nanotechnology, very realistic attractions will be easy to create and affordable. Until then, Disney and other amusement park companies will continually push available technology in order to make the background, the characters, the action, the motion, and the sounds and smells all seem more realistic. This will eventually make these attractions very believable, but it will not make the fantasies real. In other words, amusement parks will offer ever-improving virtual realities. Indeed, entertainment drives much of virtual-reality technology, because amusement parks can pay for simulations long before such equipment is affordable in the home.

Two of the most difficult effects to create are sensations of continuous extreme acceleration and weightlessness. This is because they can only be approximated by using real acceleration, which requires long distances if it is to be continuous. Thus, certain daredevil activities like jumping from airplanes cannot be easily co-opted by virtual reality—yet. Nanotechnology-based virtual reality will be able to simulate even these effects by actual continuous acceleration across the solar system, or holding one at rest away from any gravitational bodies. In any event, amusement rides will undoubtedly be a growing popular entertainment for the rest of this century and beyond.

☐ Premise and Conclusion

Whatever one's hobby, be it an old hobby in a new world, or a new hobby made possible by future changes, the abundance nanotechnology promises to provide will allow one to spend vastly more time pursuing that hobby. The machines can have dinner ready when you are done.

■ III Windows and Environments

■ 9 Phased Array Optics

Brian Wowk

By the agency of Man a new aspect of things, a new universe, comes into view.
—Francis Bacon

☐ The Future In Sight

The year is 2020. You are standing on a platform at the edge of a thousand-foot cliff overlooking the Tharsis volcano range on Mars. Your body casts a long shadow in the light of the setting sun as you scan the horizon for interesting features with your binoculars. Walking toward the edge of the precipice, you survey the texture of the rusty red boulders around you. Butterflies rise in your stomach as you peer over the cliff edge.

Sixteen hundred miles overhead, Phobos, one of the Martian moons, shines conspicuously. There is a new transport base under construction there. You move toward your telescope to get a better view when suddenly a doorway materializes out of empty space! A leg steps through it. It's your spouse with dinner.

Of course, you were never really on Mars. The "platform" you were standing on was merely the floor of a small, comfortably furnished room called a teleporter. The walls and ceiling of this room are covered with one of the technological wonders of the twenty-first century: phased array optics.

☐ Theory

Phased array optics is an application of nanotechnology that will produce three-dimensional views of objects and scenery using only two-dimensional displays. Display systems based on this technology—optical phased arrays—will behave quite literally as windows onto whatever scenery we can imagine.

Phased arrays are based on the theory of diffraction from physical optics. This theory says that patterns of light waves traveling beyond an aperture (such as a window) are entirely determined by the amplitude and phase distribution of light at the surface of the aperture. This means that if we produce light with the right phase and brightness distribution across a two-dimensional surface, we can reproduce the same light waves that would emanate from a three-dimensional scene behind the surface. In other words, we can make the surface appear as a window onto the scene.

Diffraction theory is mathematically continuous; it is assumed that surfaces can be reduced to an infinite number of small elements radiating with different amplitude and phase. This introduces complications from the standpoint of developing a practical technology. Fortunately we can achieve effective results with a discrete array of sources, provided they are coherent and less than half a wavelength apart. The wavelength of visible light ranges from 0.4 microns (violet) to 0.7 microns (red). A two-dimensional array of programmable sources 0.2 microns apart will therefore be sufficient to reconstruct any light wave pattern we desire.

Figures 9.1 through 9.3 illustrate the reconstruction principle in two dimensions for a few simple cases. In figure 9.1 all the sources radiate with the same amplitude and phase. Their waves interfere to create a plane wave propagating normal to the surface. This is the type of wave that would be produced by a distant point source. In figure 9.2 there is a linear variation of phase among the sources down the line. The result is a plane wave propagating at an angle to the surface. By adjusting the amount of phase variation, we can "steer" this beam in any direction we please. In figure 9.3 a more complicated phase relationship is used to produce a spherical wave, creating the image of a point source near the surface.

By choosing the correct phase and amplitude distribution across the array, we can, in fact, create images of any number of points at

Figure 9.1 Waves emanating from discrete sources in a phased array interface to create a plane wave that appears to come from a distant source behind the array.

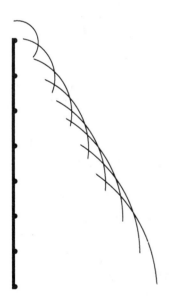

Figure 9.2 By emitting waves with different phase relationships, the array sources are able to create a plane wave that propagates at any angle to the array surface.

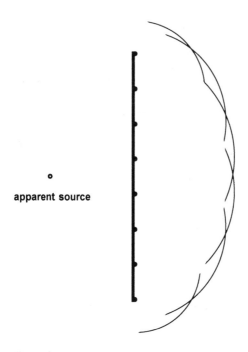

apparent source

Figure 9.3 A different combination of phase relationships produces a spherical wave that appears to come from a point near the array surface.

any number of locations behind the array. Since any three-dimensional scene can be represented as a collection of discrete points in space, it follows that an array can reproduce any three-dimensional scene.

The procedure for calculating the phase and amplitude distribution required to produce a scene is straightforward. All visible objects in the scene are represented in a computer in terms of discrete surface points. The spacing of the points is determined by the resolution of the array. Each point is assumed to produce a spherical wave. The complex amplitudes of these waves are summed at each source position on the array. The resultant complex amplitude at each source point determines the phase and intensity that source must radiate to reproduce the scene.

□ **Design**

The sources in a phased array must radiate coherently. That is, they must be able to interfere with each other. The easiest way to achieve this is to illuminate the back of the array with light from a single laser. A diverging laser beam could be aimed at the array from behind, or the beam could be transported through a thin planar waveguide on the rear surface of the array. Since lasers can be made with coherence lengths of kilometers, it doesn't really matter how the laser light gets to the sources. It will always be coherent across the array, and each source can have its own phase and amplitude calibration factors.

In this scheme, each source in the array is a passive transmission element with adjustable optical path length (phase) and transmission (amplitude). Figure 9.4 shows a cross section through a few elements. The elements are 0.2 microns (200 nm) apart. (A bacterium could cover five of them.) Each element consists of a phase shifter and an amplitude modulator.

The amplitude modulator is simple. A cross polarizer like those in LCD displays should suffice.

The phase shifter is made of an *electrooptic birefringent* material. This means that the shifter changes its refractive index along one polarization direction in response to an electric field. A polarizing filter down the back of the array ensures light of the proper linear polarization enters the shifter. A quarter-wave plate on the front of the array to restore elliptical polarization is optional.

A large variety of crystals are known to exhibit the required electrooptic effect.[1] Some of these are semiconductors that are the stock in trade of existing microfabrication technology. Unfortunately the electrooptic effect in solid crystals is rather weak. Even very large electric fields produce very small changes in refractive index. Since each phase shifter must be able to retard light passing through it by one full wavelength, a small change in refractive index means a long phase shifter. For solid crystals this length will be on the order of a millimeter, which is probably too long to be practical.

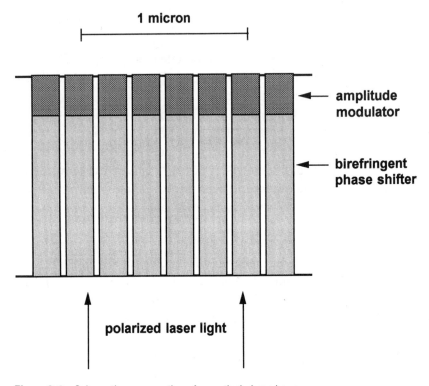

Figure 9.4   Schematic cross section of an optical phased array.

Nematic liquid crystals exhibit an electrooptic effect thousands of times stronger than that found in solids.[2] If these crystals can be made with a high refractive index and fast response, then phase shifters one micron long, such as those in figure 9.2, should be possible.

An interesting alternative has been suggested by Jeffrey Soreff of IBM.[3] Rather than relying on electrooptic effects, a crystal with a large fixed birefringence, such as calcite, could be used. The crystal is sandwiched between parallel linear polarizers and rotated relative to the direction of polarization. The phase-shifting rotation could be accomplished by mechanical rotation of the crystal or by rotating the polarizer directions. This scheme would also permit micron-length phase shifters.

In all these designs each phase shifter is behaving as a waveguide near cutoff. Red light, with a wavelength in free space of 0.6 microns, will be the most difficult to deal with. To get through a waveguide less than 0.2 microns wide, the wavelength will have to be reduced to less than 0.4 microns. This will require a refractive index of at least 1.5 in the phase shifting medium.

☐ **Data Management**

Although the synthesis of array images may be mathematically simple, it will be computationally severe (by today's standards). A source separation of 0.2 microns gives 25 million sources per square millimeter, 2.5 billion per square centimeter, and 25 trillion array sources per square meter. Each one of these sources will have to radiate with its own calculated phase and amplitude.

This is a staggering amount of information. To understand the origin of this information, recall that the array behaves as a window onto the scene it is reproducing. Suppose a square meter array is reproducing a wilderness scene. Looking at (through) the array, one sees a meadow, a small lake, and a mountain beyond. One mile away, on the far shore of the lake, a man is fishing. If you aimed a telescope in his direction, you would see him clearly. And if you had a really big telescope (one with a two-foot wide lens), you would be able to see a housefly on his hat.

It is now obvious why the array contains so much information. A meter wide array (or window for that matter) has a diffraction limited resolution of about one millionth of a radian, or 0.2 seconds of arc. The array image therefore contains every scene element subtending a solid angle greater than a trillionth of a steradian: the fly on the man's hat, leaves on trees five miles away, almost every blade of grass in the meadow, and so on.

The number of scene elements recorded by an array is roughly equal to the number of sources in the array. To compute the phase and amplitude for each source, we must add the contributions of spherical waves from each scene element to each source point. For a square meter array, this is 1000 trillion trillion floating point operations. A single computer would have trouble finishing this calculation within the lifetime of the universe.

Fortunately several considerations come to our rescue. First, this type of calculation is amenable to Fast Fourier methods, which tend to reduce N-squared operations to $N\log N$ operations (a trillionfold decrease in our case). Second, this problem is well suited to massive parallelism. With enough processors, there is no reason why array image synthesis couldn't proceed at movie frame rates. Finally, not all array images will require resolution beyond that of the unaided eye. This concession alone reduces the computation and bandwidth requirements by several orders of magnitude.

Real-time arrays still pose formidable data storage and transmission problems. A square meter array operated at human eye resolution will require about a gigabyte per frame, or 50 gigabytes per second. Interestingly, light itself is ideally suited to transmission at this rate. Modulated light could be used for both long distance transmission and local control of array sources. Instead of trillions of tiny wires, light traveling through the same waveguide as the primary laser could carry image information to decoders at periodic intervals in the array.

The data storage requirements of arrays cannot yet be met economically. (A phased array movie would consume several hundred terabytes!) Nevertheless, storage technologies able to handle arrays are on the horizon. Certainly molecular technology, with its promise of one gigabyte per cubic micron, will be more than sufficient.

□ **Phased Arrays vs. Holography**

Phased arrays and holography are both methods of wavefront reconstruction, and both can produce three-dimensional images. They differ in some important ways, however.

Holography avoids the computational requirement of arrays with a simple and elegant solution. The scene to be reproduced is illuminated with a laser. Photographic film is placed near the scene to capture reflected light (the "object" beam). At the same time, a reference beam from the same laser is shone on the film. Because they come from the same laser, the object and reference beams are coherent and produce a microscopic interference pattern that is recorded on the film.

This interference pattern is analogous to the phase and amplitude modulation of sources in a phased array. When a laser beam is shone on the developed film, the transparent areas of the interference pattern select out those parts of the beam that have the correct phase to reconstruct the recorded scene.

Holography is simple and works without a computer; however, it has inherent drawbacks. Holographic recording produces phase modulation that is too coarse for unambiguous wave reconstruction. When you shine a laser on a hologram, three beams will always be emited: the reconstructed object beam, the (unwanted) conjugate beam, and the laser beam itself. This happens because of the wide spacing of interference fringes produced by holographic recording. (The three beams are actually three orders of diffraction.) Phased arrays modulate phase and amplitude at half-wavelength intervals. This is close enough to generate only one interference maximum, which contains the reconstructed beam and nothing else.

☐ **Simulated Incoherence**

Phased arrays reproduce scenery using coherent light. This raises certain problems that have been glossed over so far. It was assumed earlier that different parts of a scene could be reproduced independently. This is not quite true for coherent light. Light waves from different points in a coherent scene would interfere with each other, generating unnatural effects such as laser speckle.

A simple solution exists. Numerous separate versions of a scene could be synthesized, each assigning a different random phase factor to points in the scene. When presented in rapid succession (say twenty versions per second), all interference effects would blur out, creating an effectively incoherent scene. In terms of eliminating laser speckle, this process is equivalent to rapidly altering the microscopic structure of a scene surface.

☐ Color

Producing colored scenes is easy. Three versions of each scene would be synthesized in red, green, and blue light (the three primary colors of human vision). The array would cycle through them rapidly, alternating between red, green, and blue lasers. Because scenes would tend to split apart into separate colors if you moved your eyes fast, you would want a fast cycle time (ideally less than a millisecond).

Although they would appear perfectly authentic to human eyes, scenes produced in this way would be easily detectable by a prism. They might not fool animals either. The ultimate solution is to cycle very quickly through the entire visual spectrum at random increments. Such an array would be difficult to detect.

☐ Transparent Arrays

If a transparent planar waveguide is used to carry laser light across the back of a phased array, then the array will be semitransparent with perhaps 30 percent transmittance (like tinted glass). This raises all sorts of interesting possibilities. For one, the array could superpose images on real scenery behind it. (A glide path for a landing aircraft is a nice example.) More profound, though, is what the array could do to the scenery itself.

In the theory of diffraction, almost every element of an optical system can be modeled as a two-dimensional surface with a complex transmission function. (A lens, for example, is equivalent to a flat surface with particular phase changing properties.) Phased arrays will turn this mathematical abstraction into technological reality. A single transparent array could behave as a programmable phase plate, Fresnel zone plate, hologram, prism, diffraction grating, or variable focal length, low f number, diffraction-limited lens. All this from an array microns thick and able to change roles in a microsecond. You could hold such a thing in your hand and swear it was from another world.

☐ The Recording Problem

For many applications the scenery produced by an array will originate entirely in a computer. When it doesn't, a way must exist to record real scenery that allows reproduction by an array. In the general case of an incoherent (naturally illuminated) scene, the only way to do this is brute force: record conventional flat images from every angle you can get.

For low-resolution reproduction, a "fly's eye" lattice of one-centimeter wide lenses could be used to record multiple views of a scene. A computer would then infer the three-dimensional structure of the scene from these views and proceed with image synthesis by methods already discussed. This system would have adequate resolution for viewing with the unaided eye. Higher resolution would require larger lenses.

Transparent arrays are well suited to this application. A transparent array could behave as a fly's eye lens to record nearby objects and later become a single huge lens to record details miles away. This type of system could quickly gather all the information needed to reproduce scenes with telescopic resolution.

☐ High-Power Arrays

The array designs presented so far operate by modulating laser light passing through them. The "sources" don't actually produce light themselves. For some applications, particularly outdoor use, a higher power design is desirable. The visual intensity of the sun (about 500 watts per square meter) is the regime we seek.

Consider putting a tiny laser at the base of each phase shifter. (Quantum well semiconductor lasers would be ideal.) These lasers provide a coherent boost to light passing through the array, essentially transforming the whole array into a resonant cavity. In other words, the array becomes one big laser with adjustable emission phase across its surface. Color could be dealt with by making the lasers tunable, or by laying them out in a red-green-blue hexagonal lattice.

Now consider a building covered on all sides by a high-power phased array. By day it could produce an image of a landscape contiguous with its surroundings, thereby rendering itself effectively invisible. By night it could do more of the same, or, for variety, create the image of a floodlit building. Perhaps it would supplement this scene with the image of a powerful, sweeping searchlight to detect and/or intimidate intruders. For those occasions when a searchlight is not intimidating enough, the phase of a few array sources could be matched to the coordinates of an intruder, with interesting consequences. Suppose an observer looking at the array sees one square meter worth of sources radiating at exactly the same phase relative to him. The sources will interfere constructively, depositing their entire output energy to a very small point at his location. In other words, the observer will be at the focus of a one-kilowatt laser (not a very healthy place to be).

Of course the same principle can be employed over the entire surface area of the building, encompassing hundreds of kilowatts of laser power. Our camouflage, decorative lighting system is now a long-range missile defense system, or directional transmitter of interstellar range. (With a beam divergence of less than a millionth of a degree, a stable 100-meter wide array would appear 1 percent as bright as the sun at any distance.)

While this example is a bit extreme, it illustrates that very small coherent sources can produce very big laser power. It follows that generating images of lasers and laser light will be an important application of phased arrays (particularly in view of the high efficiency of semiconductor lasers).

☐ **Invisibility Suits**

Displaying still scenery matching your surroundings is one form of invisibility. We could call this *passive invisibility*. With a view toward very fast and compact computers, an even more audacious possibility exists: active invisibility.

Active invisibility means an array would acquire and synthesize contiguous scenery in real time. This could be accomplished with a system a few millimeters thick. The interior layer of our hypothetical "invisibility suit" consists of photosensors comparable to a human retina. The exterior is a phased array with phase shifters of extra close spacing and high refractive index. Half of the shifters are dedicated to image production, while the other half transmit light through to the photosensitive layer. The transmitting shifters are adjusted to form a multitude of focused fly's eye images on the photosensitive layer. The system is thus able to both produce and view scenery from all angles at the same time. A powerful computer network would manage the scenery synthesis and other details, such as how suit movement affected phase relationships. Other designs are possible.

Such speculations lie at the limits of foreseeable technology. It is nonetheless amazing that something as fantastic as invisibility could be achieved using known science.[4]

☐ Front-Projection Images

Phased arrays can produce images of objects anywhere in space behind them. But that is not all they can do. With certain restrictions, they can also make objects appear *in front* of them.

Figure 9.5 illustrates the principle for a point source. By producing converging spherical waves, the array creates the image of a point source in front of it. Once again, creating a whole object is just a matter of assembling the points that constitute it. By controlling the areas of the array that reproduce various points, hidden lines and surfaces are removed, and even front-projected objects will appear solid and opaque.

Imagine you are back in your teleporter room enjoying the Martian panorama. This time the floor is also covered with phased array optics. Martian boulders now appear at your feet, among the furniture, and in one case right in front of your nose! Now that's realism.

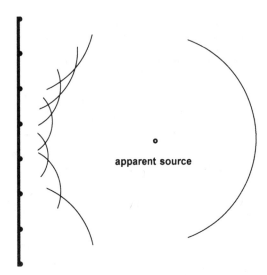

Figure 9.5 Phased arrays can produce wavefronts that appear to originate in empty space in front of the array.

Virtual teleporters covered on all sides by phased arrays bear a striking similarity to the "holodeck" concept of *Star Trek.* Anything could happen in such a room. Anything.

☐ **Eyeing The Future**

Optical phased arrays pose substantial challenges for information processing and nanofabrication. Nevertheless it seems certain that these challenges will be met within the next twenty to thirty years. Some precursor technologies, such as LCD holography, are at the edge of current capability.

Over the long term, the future of arrays is as limitless as imagination itself. The few applications mentioned here are just a glimpse of their potential. As Arthur C. Clarke said about virtual reality, this technology "won't merely replace TV. It will eat it alive."[5]

10   Utility Fog: The Stuff That Dreams Are Made Of

J. Storrs Hall

This is a most majestic vision, and Harmonious ... which by mine art I have call'd to enact my present fancies.
—William Shakespeare

An engineer of the ancient world would have gone slowly mad, trying to understand how a solid roadway could be fixed at both ends while its center traveled at a hundred miles an hour.
—Arthur C. Clarke

This that you see is simply my projection, composed of forces for which you have no name in your language.
—E. E. Smith

Sufficiently Advanced

Somewhere in your house there is a refrigerator, and within it stands a glass bottle of a beverage you wish to drink. Without getting up, or putting down this book, beckon with your hand. The door of the refrigerator will open, and the bottle will emerge, floating in the air. Simultaneously, a crystal goblet will appear in your hand. As the bottle approaches, the stopper will remove itself with a minimum of fuss. Enjoy your refreshment while the bottle floats serenely back to its place.

You decide to relandscape your yard. Go outside; somehow you become 50 feet tall, and the area looks like a model of itself. To your tools, the ground seems to have the consistency of whipped cream; but no matter how careless your step, your feet and knees aren't able to damage anything. After you've swapped the pond and the tennis court, utter a few odd-sounding words. You're normal size, the better to judge the effect. A few more magic words, like "plinth" and "colonnade," and a small Grecian temple—marble columns and all—appears. Wave your arms and it expands and contracts like some outsized accordion. When it's just right, lock it into place with a final command.

Any sufficiently advanced technology, wrote Arthur Clarke, is indistinguishable from magic. The particular invention this chapter is about, "utility fog," brings this point home poignantly. Magic is what you get when the rules of the mind describe the real, physical world. Objects appear and disappear, like thoughts do. Similarity and contagion—the "laws" of magic—are really associative retrieval and processing. With utility fog, the physical world *is* mind, in an objective literal sense that will become clear in the next few pages.

☐ A Pail of Air

Imagine a microscopic robot, about the size of a single human cell. It has a body about the size of a bacterium and several arms sticking out in all directions. A bucketful of such robots might form a "robot crystal" by linking their arms together into a lattice structure. Now take a room, with people, furniture, and other objects in it—it's still mostly empty air—and fill it completely full of robots.

With the right programming, the robots can exert any force in any direction on the surface of any object. They can hold an object up, so that it apparently floats in the air. They can hold you up, applying the same pressures to the seat of your pants that your chair would. They can exert the same resisting forces that your elbows and fingertips would expect from the arms and back of the chair. A program running in utility fog can thus simulate the physical existence of almost any object.

Utility fog operates in two modes: first, the "naive" mode where the robots act much like cells, and each robot occupies a particular position and does a particular function in a given object. In the second, or "fog" mode, the robots act more like the pixels on a TV screen. Objects are then formed of *patterns* of robots, which vary their properties according to which part of the object they are representing at the time. An object can then move through a cloud of robots without moving the individual robots, just as the pixels on your TV stay put while pictures move around on the screen.

Although each of the robots is too small to see individually, masses of them would scatter light so much as to be opaque (just like fog). The robots at the surface of a naive-mode object could hold a layer of surface plates of any desired color. Unused plates could be distributed throughout the volume of the object and physically swapped into place about as fast as a television changes its picture. In fog mode, the object would have to hold up holographic "eyephones," in front of your eyes, similar to those described elsewhere in this volume.[1]

Utility fog that is pretending to be air needs to be impalpable. You would like to be able to walk through your fog-filled house without having the feeling you had been cast into a solid block of Lucite. You'd also like to be able to breathe. To this end, the robots representing empty space constantly run a fluid-flow simulation of what the air would be doing if the robots weren't there. Each robot does what the air it displaces would do in its absence. How do you breathe when the air is a solid mass of machines? Actually, it isn't solid—foglets only take up about 5 percent of the volume they occupy (they need lots of "elbow room" to move around). Like cotton candy, they carry the air in a given volume along when the fog moves, but relatively little force is required to compress the air out when needed. Thus when fog simulates the normal air currents caused by breathing and other motions, air is moved and mixed just as it normally would be; when you inhale, fog squeezes out the appropriate amount of air.

Carried along in the fog, every six inches or so, is a pea-sized object that acts as a control center and local power reservoir. In the case of a catastrophic system failure, all the fog that is part of a naive mode object simply freezes into place, and that which is pretending to be air contracts around its local centers. In a typical room, the floor would be covered to a depth of half a foot in objects very similar to golf balls.

To understand why you'd want to fill the air with microscopic robots only to go to so much trouble to make it seem like they weren't there, consider the advantages of a TV or computer screen over an ordinary photograph. Objects on the screen can appear and disappear at will; they are not constrained by the laws of physics. The whole scene can shift instantly from one apparent locale to another. Completely imaginary constructions, impossible to build in reality, can be commonplace. Virtually anything you can imagine could be given tangible reality in a utility fog environment.

Why not, instead, build a virtual-reality machine that fools all your senses into a purely simulated version of the same apparent world? Well, in a VR machine you could experience digging up your backyard as a fifty-foot giant, but the backyard itself would miss all the fun. Fog acts as a continuous bridge between actual physical reality and virtual reality. Fog is a universal effector as well as a universal sensor. Any (real) object in a fog environment can be manipulated with an extremely wide array of patterns of pressure, force, and support; illuminated with any pattern of light; measured, analyzed, smelled, tasted, weighed, cut, reassembled, or reduced to bacteria-sized pieces and sorted for recycling.

☐ The Roads Must Roll

As well as forming an extension of the senses and muscles of individual people, fog can act as a generalized infrastructure for society at large. Fog City need have no permanent buildings of concrete, no roads of asphalt, no cars, trucks, or buses. It can look like a park, or a forest, or if the population is sufficiently whimsical, ancient Rome one day and the Emerald City the next.

It will be more efficient to build dedicated machines for long distance energy and information propagation and physical transport. Fog is ideal for local use and as an interface to worldwide communication networks. It can act as shelter, clothing, telephone, computer, and automobile. It can be almost any common household object, appearing from nowhere when you need it (and disappearing when you don't). It gains a certain efficiency from this extreme polymorphism; the same fog that was your clothing becomes your bath water and then your bed.

Another item of infrastructure that will become increasingly important in the future is information processing. Nanotechnology will allow us to build some really monster computers. Although each foglet will possess a comparatively small processor—which is to say, it will have the power of a current-day supercomputer—there are over 16 billion foglets to a cubic inch. When those foglets are not doing anything else, that is, when they are simulating the interior of a solid object or air that nothing is passing through at the moment, you could use them to balance your checkbook. Or whatever.

☐ What Computers Can't Do

When discussing something as far outside of everyday experience as utility fog, it is a good idea to delineate both sides of the boundary. Because fog is capable of so many amazing things, we should point out a few of the things it will be incapable of simulating.

▪ Anything requiring really hard surfaces, such as knives, axes, drill bits, nails, and so on. Individual foglets are hard, but then so are individual grains of confectioner's sugar. Naive-mode fog objects can be about as hard as plastic. Fog-mode fog can simulate much harder objects, but only by fooling your senses in a number of ways.
▪ Anything requiring high density. Naive-mode objects can be about as heavy as wood. Fog-mode objects can seem heavier by dynamically supplying the forces you'd feel from weight and inertia.
▪ Anything requiring both high strength and low volume. You couldn't make a parachute out of fog. (Unless, of course, all the air were filled with it, and then you wouldn't need one.)
▪ Anything requiring high heat. A fog fire could merrily blaze away on fog logs in your fireplace, and it would feel warm on your skin a

few feet away. It would feel just the same if you put your hand into the "flame."

- Anything requiring molecular manipulation or chemical transformation. Foglets are simply on the wrong scale to play with atoms. In particular, they cannot reproduce themselves. On the other hand, they can do things like prepare food the same way a human cook does—by mixing, stirring, and using special-purpose tools designed for them to use.
- Fog cannot simulate food, or anything else destined to be broken down chemically. Eating fog would be like eating the same amount of sand or sawdust.

☐ Monsters from the Id

In 1611 William Shakespeare wrote his final play, *The Tempest.* Three hundred forty-five years later a pair of obscure writers, Irving Block and Allen Adler, updated the *Tempest*'s plot into a story called "Forbidden Planet" and created a modern myth.

"Forbidden Planet," more precisely the movie version developed by Cyril Hume, has become a classic cautionary tale for any scenario in which people become too powerful and control their environment too easily. In the story, the Krell are an ancient, wise, and highly advanced civilization. They perfect an enormous and powerful machine, capable of projecting objects and forces anywhere in any form, following the mental commands of any Krell. The machine works "not wisely but too well," manifesting all the deeply buried subconscious desires of the Krells to destroy each other.

Utility fog will provide humans with powers that approximate those of the fictional Krell machine. Fortunately, we have several centuries of literary tradition to guide us around the pitfalls of hubris made reality. We must study this tradition, or we may be doomed to repeat it—a truth that is by no means limited to utility fog, or indeed to nanotechnology in general.

The first thing we can do is to require fully conscious, unequivocal commands in order for the fog to take any action. Beyond that, we can try to suggest some of the protocols that may be useful in managing the fog in situations where humans are interacting in close physical proximity. Even if we solve the problem of translating one's individual wishes, however expressed, into the quadrillions of sets of instructions sent to individual foglets to accomplish what one desires, the problem of who gets to control which foglets is probably a much more contentious one.

We can, if we desire, physicalize the psychological concept of "personal space." Foglets within some distance of each person would be under that person's exclusive control; personal spaces could not merge except by mutual consent. This single protocol could prevent most crimes of violence in our hypothetical Fog City.

A corollary point is that physically perpetrated theft would be impossible in a fog world. It would still be possible by informational means—fraud, cons, hacking, and so on—but fog could be programmed to put ownership on the level of a physical law (not that it makes much sense to think of stealing a fog-mode object, anyway). Ownership and control of fog would not be greatly different in principle from ownership and control of land. Most of us are used to such laws, but it's actually a complicated bundle of rights: you don't own the road, but you have exclusive rights to the part you're driving on right now. You do own your car, but you will discover limits if you park in a tow-away zone.

Indeed, much of the fog's programming will need to have the character of physical laws. In order for the enormous potential complexity to be comprehensible and thus usable to human beings, it needs to be organized by simple but powerful principles that are consonant with the immense, hard-wired information processing that our sensory systems perform. For example, it would be easy to move furniture (or buildings) by manipulating a scale model with your hands, and easy to see what's going on by looking at the model. However, fog could just as easily have flooded the room with 100 kHz sound, and frequency-scaled the echoes down into your auditory range. A bat would have no trouble with this kind of "scale model," but to us it's just noise.

It will be necessary, in general, to arrange the overall control of the fog to be extremely distributed, as local as possible, and robust in the presence of failure. When we realize that a single cubic inch of fog represents a computer network of greater complexity and computing capacity than any existing in the world today, the concept of hierarchical control with human oversight can be seen to be hopelessly inadequate. Agoric distributed-control algorithms that barter locally for computational resources offer one possible solution.[2]

☐ **With Folded Hands**

Another major advantage for space-filling fog is safety. In your car (or its nanotech descendant) fog forms a dynamic form-fitting cushion that protects you better than any nylon-fiber seat belt. An appropriately built house filled with fog could even protect its inhabitants from the effects of a nuclear weapon—within 95 percent or so of its lethal blast area.

There are many more mundane ways fog can protect its occupants, not the least being to physically remove bacteria, mites, pollen, and so forth, from the air. A fog-filled home would no longer be the place where most accidents happen. First, by performing most household tasks using fog as an instrumentality, the cuts and falls that accompany the use of knives, power tools, ladders, and so forth can be eliminated.

Second, the other major class of household accidents, injuries to young children who hurt themselves out of ignorance, can be avoided by a number of means. A child who climbed over a stair rail would float harmlessly to the floor. A child could not pull a bookcase over on himself; falling over would not be among the bookcase's repertoire. Power tools, kitchen implements, and cleaning chemicals would not normally exist; they or their analogues would be called into existence only when needed and then vanish, instead of having to be cleaned and put away.

Outside the home, the possibilities are, if anything, greater. One can easily imagine an "industrial fog" factory. It would consist of larger robots, perhaps the size of grains of sand. Of course no one would care that it couldn't simulate the textures and appearances of silk or fine china! Unlike domestic fog, which would have the density and strength of balsa wood, industrial fog could have bulk properties resembling hardwood or aluminum. A nanotechnological factory would probably consist of a mass of fog with special-purpose reactors embedded in it, where high-energy chemical transformations could take place. All the physical manipulation, transport, assembly, and so forth would be done by the fog—providing the ultimate in flexible manufacturing.

☐ **From Earth to Moon**

The major components and systems of spaceships will need to be made with special-purpose nanotechnological mechanisms, and indeed with such mechanisms pushed much closer to their true capacities than anything we have mentioned heretofore. In the cabin of your spaceship, however, you will want an acceleration couch, a fine application for fog. But when you're not accelerating, which is most of the time, you'd prefer there to be something useful, like empty space.

Put utility fog in the cabin and you never have to worry about floating out of reach of a handhold. Instruments, consoles, and cabinets for equipment and supplies are not needed. Items that cannot be simulated can be embedded in the fog in what are apparently bulkheads.

Fog can add great structural strength to the ship itself; the rest of the structure need be not much more than a balloon. The same is true with your spacesuit: fog inside the suit manages the air pressure and makes motion easy; fog outside gives you extremely fine manipulating ability for various tasks. Of course, like the ship, the suit contains many special-purpose, nonfog mechanisms.

Surround your space station with fog; it needs radiation shielding anyway. These would be big industrial foglets with lots of redundancy in the mechanism, and even so they may get recycled fairly often. All the stock problems from science fiction movies go away: you never need to go outside to fix something; and if you *want* to go outside, you can't let go and go spinning off into space. Outside, fog also makes a good tugboat for docking spaceships.

When you homestead your little patch of the Moon, bring along a batch of heavy duty fog as well as special-purpose nanotech power generation and waste recycling equipment. There will be a million and one tasks, of the ordinary yet arduous physical kind, that must be done to set up and maintain a self-sufficient household.

☐ **The Shape of Things to Come**

Virtually every physical job humans do could be performed by utility fog. Most nonphysical tasks—those involving communication and decision making—could be greatly facilitated by fog-mediated telepresence and computational power. (The question of whether artificial intelligence could replace those functions entirely is beyond the scope of this chapter.)

The appearance and functionality of each individual's environment will be completely programmable, and we should expect an extremely wide variety of styles to emerge. However, many of the less lovely aspects of life, such as dirty crowded cities, which are products of necessity, could reasonably be expected to decrease.

To demonstrate how fog would appear to its inhabitants and explain what would actually be happening underneath, let us consider the following vignette. Each statement of a surface effect is followed by a description of how the effect is implemented by the fog.

"I decided to see if Fred was interested in breakfast. He was in, so I popped over."

(Fred lives 100 miles away. You subvocalized a command to inquire about his status. Utility fog robots in a layer next to your skin, acting as EEG receptors, are able to read enough of your nerve activity to interpret this command. A communication pathway is opened between your domicile and Fred's; this consists of a foglet-to-foglet network between you and the nearest hard-matter data transmission

node, and along that network to whatever batch of fog Fred has act-
ing as his "social secretary" at the moment. Fred's program knows
that he is accepting visitors, at least you, at the moment, so an unoc-
cupied section of fog in Fred's place is bound over to your control. A
telepresence link is set up; the fog assumes your shape and transmits
sights, sounds, and forces to the fog around you, which assumes the
appearance of the remote location. To your senses and Fred's, you
have been instantly transported to his location.)

"Fred looked like the very devil, and in fact his place was a pretty
decent Hell. Carefully avoiding a pit of bubbling lava, I decided the
only appropriate shape for me was Frosty the Snowman."

Assuming that the appearance and physical properties of a shape
are programmed, the software techniques to simulate it are straight-
forward extensions of current-day computer graphics. Fog has more
than adequate computational power to do this in real time. Fog
around a person can easily simulate an elaborate costume, using, if
desired, telepresence techniques to make the person feel as if he ac-
tually had taken on that shape.

"We settled on La Belle Maison for breakfast on the way to the
zoo."

The zoo is a real place, since people like to see animals in the
flesh. Both of you must travel there physically. Each of you will travel
in a capsule that simulates the interior of a restaurant, including a
telepresence of the other. The restaurant itself is on the other side of
the world. In response to your query (and cash!), it downloads the
appropriate software for simulating its interior to both of you.

"Gaston's crepes were magnificent, as always. I enjoy the Old
World massiveness of La Belle Maison's furnishings, especially
those solid silver utensils."

Food, the most common item fog cannot simulate, is assembled
by "standard" molecular construction under the direction of other
programs downloaded from the restaurant. The developing food
forms a tree of coalescing subparts that intercepts your transporta-
tion trajectory just as the waiter, who is a program rather than a
telepresence, appears to bring it from the kitchen. Even though the

substance of fog isn't much denser than balsa wood, it can appear to have weight and mass by simulating the forces that gravity and inertia would cause it to exert against your hand. It can do this as long as any part of the object is in contact with the "air."

"As we wandered around the zoo, a family of leopards took an interest in us and followed us around for a while."

Physicalized "personal space" could easily protect people and animals from each other in a zoo, allowing them to intermix freely without cages in perfect safety. For highly nervous animals, it would be possible to program invisibility between them and their predators, or for the visitors if they would overly excite the animals.

"Then we flew over to the jungle area. The monkeys chattered at us from the high branches."

Flying individually is no different than using fog to move any other object. It feels, presumably, more or less like swimming. You would almost certainly want to map the control for attitude, speed, and direction into motor impulses instead of verbal commands; it's actually a simple technology, used today for prosthetics.

"Afterward, we zipped on up to the promenade to check out the latest fashions."

In the vicinity of the floating city, your trajectory is more closely controlled, so you generally request a least-time routing. The city is a mass of structures of varying permanence; there is no need for streets or walkways, but mall-like promenades are provided in some commercial sectors for that quaint twentieth-century ambiance.

☐ **Rossum's Universal Robots**

Compared with an assembler, a foglet will be huge and overpowered, able to control its motions to a tenth of a micron instead of a tenth of a nanometer. It will have an arm-spread of ten microns. Each foglet will weigh about 20 nanograms and contain about 5 trillion atoms.

This is about as small as it would be feasible to make a foglet. There is no obvious upper bound on size, except to reduce the resolution and verisimilitude of the simulation. Foglets whose appearance was unimportant and that were simply to manipulate objects could be on the order of inches or even feet. It would probably be workable to have foglets ten or even a hundred times as large as the design presented here, which would simplify some of the engineering problems. They would be visible to the naked eye, if you looked closely, but then so are the pixels on your television.

Most currently proposed nanotechnological designs are based on carbon. Carbon is a marvelous atom for structural purposes, forming a crystal (diamond) that is very stiff and strong. However, a fog built of diamond would have a problem that nanomechanical designs of a more conventional form do not pose: fog has so much surface area exposed to the air that if it were largely diamond, especially on the surface, it would amount to an easily ignitable fuel-air explosive.

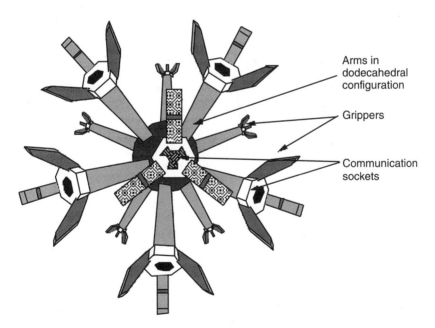

Arms in
dodecahedral
configuration

Grippers

Communication
sockets

Figure 10.1 A foglet, containing some five trillion atoms, with a reach of about ten microns.

Therefore, the basic foglet is designed so that its structural elements, forming the major component of its mass, are made of aluminum oxide, a refractory compound using common elements. The structural elements form an exoskeleton, which, besides being a good mechanical design, allows us to have an evacuated interior in which more sensitive nanomechanical components can operate. Of course, any macroscopic ignition source would vaporize a foglet entirely; but as long as more energy is used vaporizing the exoskeleton than is gained burning the carbon-based components inside, the reaction cannot spread.

Each foglet has twelve arms, arranged as the faces of a dodecahedron. The arms have telescoping mechanisms rather than joints. Each arm swivels on a universal joint at the base, and the gripper at the end can rotate about the arm's axis. Each arm thus has four degrees of freedom, plus opening and closing the gripper. The swivel and rotate axes are weakly driven, able to position the arm in free air but not drive any kind of load. The only load-carrying motor on each axis is the extension/retraction motor.

The gripper is a hexagonal structure with three fingers mounted on alternating faces of the hexagon. Two foglets "clasp hands" in an interleaved six-finger grip. Since the fingers are designed to match the end of the other arm, this provides a relatively rigid connection; forces are only transmitted axially through the grip.

When at rest, foglets form a regular lattice structure. If the bodies of the foglets are thought of as atoms, it is a *face-centered cubic* crystal formation, in which each atom touches twelve other atoms. If you imagine the foglets' arms as the girders of a trusswork bridge, they form a configuration known as an octet truss, invented by Buckminster Fuller in 1956.[3] The spaces bounded by the arms form alternate tetrahedrons and octahedrons, both of which are rigid, three-dimensional shapes.

A mass of fog may be thought of as consisting of layers of foglets. The layers, and the shear planes they define, lie at four major angles (corresponding to the faces of the tetrahedrons and octahedrons) and three minor ones (corresponding to the face-centered cube faces). In each of the four major orientations, each foglet uses six arms to hold its neighbors in the layer; layers are thus a two-dimensionally rigid fabric of equilateral triangles. In face-centered mode, the layers work out to be square grids and are thus not rigid, a slight disadvantage. Most fog motion is organized in layers; layers slide by passing each other down hand-over-hand in bucket-brigade fashion. At any instant, roughly half the arms will be linked between layers when they are in motion.

Each moving layer of robots is similarly passing the next layer along, so each layer adds another increment of the velocity difference between adjacent layers. If we assume that an arm's speed is a meter per second, and the interlayer distance is an arm's length (ten microns), we get a maximum shear rate of one hundred kilometers per second per meter. This means that if Sandy Koufax were standing in solid fog, pitching a 100 mile-per-hour fastball, the boundary layer of foglets allowing his hand to pass need only be four tenths (0.447) of a millimeter thick.

The atomically precise crystals of the foglets' structural members will have a tensile strength of at least 100,000 pounds per square inch (psi), or higher than most steels but lower than modern "high-tech" composites using refractory ceramics. With an arm's length of ten microns, fog will occupy 5 percent of the volume of the air but has structural efficiency of only about 1 percent in any given direction.

Thus utility fog, as a bulk material, will have a density (specific gravity) of 0.2. For comparison, the density of balsa wood is about 0.15, cork is about 0.25. In action, fog will have a tensile strength of only 1000 psi, which is about the same as low-density polyethylene (solid, not foam). Luckily the material properties arising from the lattice structure are more or less isotropic; the one exception is that when fog is flowing, tensile strength perpendicular to the shear plane is cut roughly in half.

Without altering the lattice connectivity, fog can contract by up to about 40 percent in any linear dimension, reducing its overall volume (and increasing its density) by a factor of five. (This is of course done by retracting all arms while not letting go.) In this state the fog has the density of water. An even denser state can be attained by forming two interpenetrating lattices and retracting; at this point its density and strength would be similar to ivory or Corian structural plastic: a specific gravity of 2 and tensile strength of about 6000 psi. Such high-density fog would have the useful property of being waterproof (which ordinary fog is not), but it cannot flow and takes much longer to change configuration.

"Industrial" fog could, by use of thicker arms, increase the density and strength of ordinary fog by a factor of three, without losing mobility. In a retracted state, properly designed industrial fog could be waterproof without interpenetrating, making it more flexible in applications demanding that property.

The Mightiest Machine

For a mass of fog to press against some object at its peak 1000 psi, each foglet arm must sustain a force of about 30 dynes. For the arm to go from fully retracted to fully extended will take 12 milliergs of energy. If the foglet ran on hydrogen, it would burn about 3 billion atoms to get this much energy. That may sound like a lot, but it's less than 0.1 percent of the volume of a foglet, a pretty small fuel tank.

In fact, foglets run on electricity, but they store hydrogen as an energy buffer. Hydrogen is used in part because it's almost certain to be a fuel of choice in the nanotech world, and thus we can be sure that the process of converting hydrogen and oxygen to water and energy—as well as the process of converting energy and water to hydrogen and oxygen—will be well understood. That means we'll be able to make these conversions efficiently, which is of prime importance.

Suppose that the fog is flowing, layers sliding against each other, and some force is being transmitted through the flow. This would happen any time fog moved a nonfog object, for example. Just as your muscles oppose each other when you hold something tightly, opposing forces along different foglet arms act to hold the fog's shape and supply the required motion.

When two layers of fog move past each other, the arms between may need to move as many as 100 thousand times per second. If each of those motions were dissipative, and the fog were under full load, it would need to consume 700 kilowatts per cubic centimeter. This is roughly the power dissipation in a .45 caliber cartridge the millisecond after the trigger is pulled—which is to say: it just won't do.

But nowhere near this amount of energy is actually being used; pushing arms are supplying this much force, but the arms being pushed are *receiving* almost the same amount, minus the work being done on the object being moved. So if the motors can act as generators when they're being pushed, each foglet's energy budget is nearly balanced. Because these are arms instead of wheels, intake and outflow do not match at any given instant, even though they average out over time (measured in tens of microseconds). Some buffering is needed. Hence the hydrogen.

I should hasten to add that almost never would you expect fog to move actively at 1000 psi; even to raise your body, the pressure in the column of fog beneath you is less than one thousandth of that. The 1000 psi is available so that fog can simulate hard objects, where forces can be concentrated into very small areas. Even so, current exploratory engineering designs for electric motors have power conversion densities up to a billion watts per cubic centimeter and dissipative inefficiencies in the 10 parts per million range. This means that if the Empire State Building were being floated around on a column of fog, the fog would dissipate less than a watt per cubic centimeter.

Moving fog will dissipate energy by air turbulence and viscous drag. In the large, air will be entrained in the layers of moving fog and forced into laminar flow. Energy consumed in this regime may be properly thought of as necessary for the desired motion. As for the waving of the arms between layers, the Reynolds number decreases linearly with the size of the arm.[4] Since the absolute velocity of the arms is low, that is 1 m/s, the Reynolds number should be well below the "lower critical" value, and the arms should be operating in a perfectly viscous regime with no turbulence. The remaining effect, viscous drag (on the waving arms), comes to a few watts per square meter of shear plane per layer.

There will certainly be some waste heat generated by working fog that will need to be dissipated. This and other applications for heat pumps, such as personal heating or cooling (no need to heat the whole building, especially since people prefer different temperatures), can be accomplished simply by running a flow of fog through a pipelike volume that changes in area, compressing and expanding the entrained air at the appropriate places.

☐ **Marooned in Realtime**

In the macroscopic world, microcomputer-based controllers (e.g., the widely used Intel 8051 series microcontrollers) typically run at a clock speed of about 10 MHz. They emit control signals on the order of 10 kHz (usually less) and control motions in robots that are at most 10 Hz. Thus a complete motion takes one tenth of a second. A million clock cycles per action is not strictly necessary, of course, but it gives us some concept of the action rate we might expect for a given computer clock rate in a digitally controlled nanorobot.

Analysis shows that it is possible to build *mechanical* nanocomputers with gigahertz clock rates.[5] From this it is clear that we can build a nanocontroller that can direct a 10-KHz robot. We can do better, however.

Computer architectures have advanced since the early microcontrollers were developed. The 8051 completes a single instruction in 6, 12, or 18 clock cycles; modern RISC (reduced instruction-set computing) architectures execute one instruction each cycle. So far, nobody has bothered to build a RISC microcontroller; they already have more computing power than they need. RISC designs are also efficient when it comes to hardware; one early RISC chip was implemented on a 10,000-gate gate array. This particular design could be translated into a Drexlerian, rod-logic machine using less than one tenth of one percent of a cubic micron.

Each foglet has twelve arms with three axis controls each. In current technology it isn't uncommon to have a processor *per axis;* we could easily fit thirty-six processors into the foglet, but it isn't necessary. The tradeoffs in macroscopic robotics today are such that processors are cheap; in a foglet things are different. Control of a foglet's arms is actually much simpler than control of a macroscopic robot. They can be managed by much simpler controllers that need to execute commands like, "Move to point x at speed y." Using a RISC design allows a single processor to control a 100 kHz arm; using auxiliary controllers will let it control all twelve easily. But there is still a problem.

There is a fundamental thermodynamic law that states that every time a bit is erased in a computer, you must dissipate $kT \ln 2$ joules of energy.[6] This includes not only bits explicitly erased in a register but also ordinary logic operations such as a NOR (not-or) gate where two bits go in and only one comes out. Rod logic (or other nanotechnological computer hardware) designs are quite efficient in that they get near this physical limitation; but it's better if they do not. Ten-thousand gates of conventional logic operating at a gigahertz (at room temperature) must dissipate 27 nanowatts of heat even if they are perfectly efficient.

Twenty-seven nanowatts may not seem like much, but there are a billion foglets in a cubic centimeter. So each cubic centimeter of fog must emit 27 watts of heat—from its computers alone. One cubic meter would put out 27 megawatts; the volume of your whole house doesn't even bear thinking about. Inasmuch as this is a physical limit, some radical rethinking is necessary.

Luckily, some radical rethinking has already been done. The estimates just used for irreversible bit operations were derived from the design of a conventional macroscopic processor. In conventional design, the irreversibility of bit operations is not an issue, but we should design nanoprocessors to eliminate such operations as much as possible. Some recent work on reversible computers by Ralph Merkle, head of the Computational Nanotechnology Project at Xerox PARC, and I, show that it is possible to build computers not very different in basic instruction-level organization than conventional ones, but which only execute irreversible operations on certain instructions.[7] Such designs should reduce the cost to an average of 10 bits per cycle.

This would bring the power dissipation down to around 30 milliwatts per cubic centimeter, which is roughly the same as the dissipation required from mechanical motion! As long as the computers can go into a standby mode when the fog is standing still, this is quite workable. Concentrations of heavy work—mechanical or computing—would still require cooling circulation to a degree, but, as we have seen, fog is perfectly capable of doing that.

□ **The Ophiuchi Hotline**

What about all the other computing overhead for fog? Besides the individual control of its robotic self, each foglet will have to run a portion of the overall distributed control and communications algorithms. We can use another clock-speed-to-capability analogy from current computers regarding communications. Megahertz-speed computers find themselves well employed managing a handful of megabit data lines. Again we are forced to abandon the engineering tradeoffs of the macroscopic world: routing a message through any given node need theoretically consume only a handful of thermodynamically irreversible bit operations; typical communications controllers take millions. Special-purpose message routers designed with these facts in mind must be a part of the foglet.

If fog were configured as a store-and-forward network, using packets with an average length of 100 bytes and a 1000-instruction overhead, information could move through the fog at 5 meters per second, or 11 miles per hour. You could beat this by writing the information on a paper airplane and throwing it across the room.

This approach represents a highly inefficient use of computation even with special-purpose hardware. A more efficient communication protocol will be needed.

The essence of such a protocol will be the ability to skip over intervening foglets. There are two ways this might be done. First, long chains of linked arms could be circuit-switched into end-to-end conductivity, forming "long wires" through which a form of "wormhole" routing could be achieved. Second, if we use larger foglets, we could add optical wave-guides that ran down the foglet's arms. Larger foglets would also reduce the number of links required between any two given points.

Cellular automata-based algorithms might provide another source of efficiency that could be used for simulating materials that fog is supposed to be representing. These algorithms can be extremely parsimonious in irreversible operations, and they are ideally suited for the massively parallel, distributed computational base fog provides.

☐ **The Rest of the Robots**

The counterintuitive inefficiency in communications is an example, possibly the most extreme one, of a case where macroscopic mechanisms outperform utility fog at a specific task. Other examples will emerge when we consider nanoengineered macroscopic mechanisms.

We could imagine a robot, human-sized, that was formed of a collection of nanoengineered parts held together by a mass of industrial utility fog. The parts might include "bones," perhaps diamond-fiber composites, having great structural strength, as well as motors, power sources, and so forth. The parts would form a sort of erector set that the surrounding fog could assemble to perform the task at hand. Fog could perform directly all subtasks not requiring the excessive strength, power, and so forth that the special-purpose parts would supply.

A fog house, or city, would resemble a fog robot in this regard. The roof of a house might well be specially engineered for specific qualities—waterproofness, solar energy collection, resistance to general abuse, and so on—far exceeding that which ordinary general purpose fog would have. (On the other hand, fog could, if desired, have excellent insulating properties.) Of course the roof need not be one piece—it might be inch-square tiles held in place by supporting fog, and thus be quite amenable to incremental repair and replacement, and rearrangement at the owner's whim, and all the other advantages we expect from a fog house.

Other major components that would be profitably built from special-purpose nanomaches are power and communications modules. Working on more-efficient protocols, as suggested above, fog would form an acceptable communications link from a person to some terminal in the same building, but it would be extremely inefficient for long-haul, high-bandwidth connections such as would be required for telepresence. Power is also almost certainly the domain of special-purpose mechanisms. Power transmission in the fog is likely to be limited, although for different reasons from data transmission. Nanotechnology will give us an amazing array of power generation and distribution possibilities; fog can use most of them.

The critical heterogeneous component of fog is the fog-producing machine. Foglets are not self-reproducing; there is no need for them to be, and it would complicate their design enormously to give them fine, atom-manipulating capability. One imagines a fog generator the size of a bread box producing fog for a house, or building-sized machines filling cities with fog. Fog itself, of course, would convey raw materials back to the machine.

☐ **Tunnel in the Sky**

Fog moves an object by setting up a seed-shaped zone around it. The foglets in the zone move with the object, forming a fairing that makes the motions around it smoother. If the object is moving fast, fog in its path will compress to let it go by. The air in a fog matrix does not have time to move, so the motion is fairly efficient. For slower motions, efficiency is not so important, but if we wish to prevent slow-moving, high-pressure areas from interfering with other airflow operations, we can enclose the object's zone in a self-contained convection cell that moves foglets from in front of it to behind.

The Cold Equations

Utility fog is actually one of the simpler nanotech devices. It does not have to live between your cells, like medical nanorobots. It does not have to manipulate single atoms, like assemblers. It is physically large, so we do not have to push any theoretical design limits to get everything inside. The fog's motors, computers, and communication systems are all well within the limits of conservatively applied engineering principles.

It's a bit ironic that the hardest part of the fog is the part we can do right now: the software. To be lived in, fog needs to be very reliable. Physically, that's not too hard; a foglet that breaks down becomes a tiny speck of dust and can be cleaned out of the way like all the other specks by the remaining fog. Furthermore, any individual foglet that tried to do the wrong thing wouldn't accomplish much. But let a distributed control program get loose and all sorts of mischief could happen.

Of course, this isn't a problem for fog alone. Almost any believable scenario for future technology involves ever more complex software performing ever more important functions. Already banks, phones, air traffic control, and a host of other institutions that our lives depend on are run by complex, real-time, distributed programs. Perhaps the prospect of living physically embedded in utility fog will enhance the perceived need for simplicity, reliability, and predictability, ultimately improving the quality of all our computational systems.

Acknowledgments

The author gratefully acknowledges technical criticism and suggestions from, among countless others, K. Eric Drexler, Ralph Merkle, and Carl Feynman.

Our revels now are ended. These our actors,
(As I foretold you) were all spirits, and
Are melted into air, into thin air,
And, like the baseless fabric of this vision,
The cloud-capp'd tow'rs, the gorgeous palaces,
The solemn temples, the great globe itself,
Yea, and all which it inherit, shall dissolve
And, like this insubstantial pageant faded
Leave not a rack behind. We are such stuff
As dreams are made on, and our little life
Is rounded with a sleep.
—William Shakespeare

■ Postscript

Ebola Zaire attacks every organ and tissue in the human body except skeletal muscle and bone. It is a perfect parasite because it transforms virtually every part of the body into a digested slime of virus particles. The seven mysterious proteins that, assembled together, make up the Ebola-virus particle, work as a relentless machine, a molecular shark, and they consume the body as the virus makes copies of itself.
—Richard Preston

Superintelligence is not perfection—spectacular failures are certain. For this reason, diversity is to be desired and expected.
—Hans Moravec

In a world with Ebola and other viruses, such as HIV, a microscopic sheathing of utility fog or some such similar material might be welcome.[1] Adopting a molecular layer of flexible diamondoid skin, we may be able to protect our health as individual monads within deteriorating and increasingly dangerous ecosystems. It is, however, far from certain that radically separating the body from its environment would be viable.

Alternatively, if we can create molecular skins that would protect our flesh from microscopic viruses and macroscopic violence, it is likely that we will also be able to encapsulate entire towns or ecologies with similarly strong, flexible, and intelligent materials. Architectural descendants of Biosphere II and Buckminster Fuller's geodesic domes could be used to create a multitude of distinct, self-contained, human-bearing terrariums—terrariums that need not remain on Earth. By first growing the enclosing membrane under the enclosure to form a complete sphere, and then, pending a confidence vote by the sphere's population, opening trillions of microscopic vacuum chambers in the upper portion of the enclosure's nano-skin, a terrarium could rise up into the air, using the density of the atmosphere to buoy itself out into space, and float up to the top of the tratosphere. From there, space tugs could haul the minute world to a near-Earth solar orbit where it could join with other encapsulated townships and ecologies to regroup, coordinating themselves into spinning tubular structures, centrifugally generating a sense of gravity.

Such a transition is not entirely without precedent. When cellular life emerged from primeval pools, and, much later, when groups of cells organized themselves into composite creatures, the natural order of living systems was radically reconfigured. If the development of nanotechnology is successful, and we are able to manage a molecular transformation of human culture, we may recapitulate such processes of separation and reorganization at a still higher level of organizational complexity.

Considering the explosive growth of the human population and the consequent ecological devastation of the planet, it is not difficult to argue that the success of our species has driven the Earth out of balance. One could say, with Robert Preston, that "the earth is mounting an immune response against the human species."[2] And this response implicates the machinic infrastructure as much as the organic. Hemorrhagic viruses such as Ebola are everyone's concern because of the global network of airplane traffic. Today there is only one living experiment: the Earth. Developing the technology to launch a multitude of ecologically self-contained experiments could dramatically increase the chances for human survival. If the planet *is* intent upon clearing itself of a human fever, there appear to be only two alternatives: widespread human death or ecosystem migration. Indeed, the "purpose" of nanotechnology may be its capacity to enable the global flowering of Earth life into space.[3]

But such speculation is at best a beginning, the rosy-fingered dawn that precedes decision and action. Integrating the rising tide of molecularly precise technologies into the real-world matrix of rapidly changing human cultures will not be easy. We have yet to evolve the political or economic—or indeed the ethical or aesthetic—mechanisms that will allow humans as we have known them in the twentieth century to coexist with the molecularized products of the machinic phylum in the twenty-first. If you care to join me in an effort to evolve such mechanisms, you can find me at the Molecular Realities World Wide Web site (*http://www.well.com/~bcc/MolecularRealities.html*).

When we think about the future of the world, we always have in mind its being at the place where it would be if it continued to move as we see it moving now. We do not realize that it moves not in a straight line, but in a curve, and that its direction constantly changes.
—Ludwig Wittgenstein

■ Contributors

The most general formula of the relation to oneself is the affect of self by self, or folded force. Subjectivation is created by folding. Only, there are four foldings, four folds of subjectivation, like the rivers of the inferno.
—Gilles Deleuze

■ **Harry Chesley** is a senior software architect at Macromedia developing networked multimedia. He has worked with computer software and communications for twenty-five years, including eight years at Apple Computer and three at SRI International. His Internet address is chesley@flights.com

■ **BC Crandall** is the founder and director of Molecular Realities and the founder and president of Memetic Engineering. He is also the cofounder of Prime Arithmetics, Inc. He edited the proceedings of the first international conference on nanotechnology, *Nanotechnology: Research and Perspectives,* which was published by The MIT Press in 1992. His Internet address is: bcc@well.com

■ **Richard Crawford** worked at Lawrence Livermore National Laboratory as the senior editor of their prestigious monthly, *Energy and Technology Review,* until his retirement in 1984. Since then, he has continued to act as a consultant to Lawrence Laboratories and others. He first read about nanotechnology in 1988 and has been actively pursuing the subject ever since.

■ **J. Storrs Hall** is a computer scientist specializing in massively parallel processor architectures at the Laboratory for Computer Science Research, Rutgers University. He is founder and moderator of Usenet's "sci.nanotech" newsgroup and is the nanotechnology editor for *Extropy* magazine. Dr. Hall is married and has two robots. His Internet address is: josh@cs.rutgers.edu

- H. Keith Henson is best known for founding the L5 Society, now absorbed into the National Space Society. He is a senior member of IEEE and a consultant on hardware/software projects, most recently a CD-ROM jukebox. He holds patents in several fields, including two held jointly with K. Eric Drexler. His early contacts with Dr. Drexler led through nanotechnology to cryonics, and he is now on the board of the Alcor Life Extension Foundation. Mr. Henson has also written extensively on memetics. He can be reached on the Internet at: hkhenson@cup.portal.com

- Ted Kaehler, a computer scientist in the Advanced Technology Group at Apple Computer, Inc., received his B.S. in physics from Stanford University and an M.S. in computer science from Carnegie-Mellon University. He spent a sabbatical at the Foresight Institute and has spoken frequently on nanotechnology. Mr. Kaehler leads two discussion groups on the social and technical implications of molecular manufacturing for the Computer Professionals for Social Responsibility. His Internet address is: Ted_Kaehler@atg.apple.com

- Tom McKendree has tracked nanotechnology since reading a pre-publication copy of K. Eric Drexler's *Engines of Creation* in 1986 while at MIT. After graduating, with a B.S. in mathematics and a B.S. in aeronautics and astronautics, he entered aerospace and is now at Hughes. In 1989 he received an M.S. in systems from the University of Southern California where he is currently pursuing a Ph.D. in systems architecture. He is president of the Molecular Manufacturing Shortcut Group, which studies how nanotechnology can be used as a shortcut for developing space. Mr. McKendree can be reached on the Internet at: mckendree@chaph.usc.edu

- John Papiewski works at RMS Business Systems, a software consulting firm in the Chicago suburbs. He writes fiction and computer graphics software.

- Edward M. Reifman received his B.S. in mechanical engineering, magna cum laude, in 1971 and an M.S. in biomedical engineering in 1972, both degrees from UCLA. Upon graduation, Dr. Reifman worked at Hughes Aircraft Company designing communications satellites until 1974, when he was admitted to the UCLA School of Dentistry. He received his D.D.S. from UCLA in 1978. Upon graduation, Dr. Reifman won an honorarium to attend and lecture at the Karl Haupl Dental Institute, Dusseldorf, Germany. Dr. Reifman has a private practice in Encino, California.

▪ Brian Wowk is a graduate student at the National Research Council Institute for Biodiagnostics, Winnipeg, Canada, where he pursues research in magnetic resonance imaging. He received a master's degree in physics from the University of Manitoba in 1993 and is currently a Ph.D. candidate. His research interests include functional neuroimaging, therapy imaging, and medical imaging in general. Other interests are aging research, molecular engineering, and the medical applications of nanoscale technologies. He is a member of the American Association of Physicists in Medicine and the Society of Magnetic Resonance. Mr. Wowk can be reached on the Internet at: wowk@ccu.umanitoba.ca

■ Notes

☐ Preface

The epigraph is from Valentino Braitenberg, *Vehicles: Experiments in Synthetic Psychology* (Cambridge: The MIT Press, 1984), 1.

1. Ed Regis, *Nano: the Emerging Science of Nanotechnology: Remaking the World—Molecule by Molecule* (Boston: Little, Brown and Company, 1995).

2. Markus Krummenacker and James Lewis, *Prospects in Nanotechnology: Toward Molecular Manufacturing* (New York: John Wiley & Sons, 1995). See also Max Nelson and Calvin Shipbaugh, *The Potential of Nanotechnology for Molecular Manufacturing* (Santa Monica, CA: The Rand Corporation, 1995).

3. The flyer I distributed at the Palo Alto conference on nanotechnology in November of 1991 read as follows:

Are Molecules Sacred?

Are molecules sacred? If not, what is?—if anything? What is worthy of deep respect and honor? Perhaps life. But what is life other than temporarily existing molecular configurations with a limited capacity to carry certain self-reproducing patterns of information (genes, memes). When molecular configurations as complex and as capable of transporting and transforming informational patterns can be designed and engineered by humans, will these patterns be sacred?

We are swirling molecular systems, as is the earth, the moon, and the stars. Long ago the Buddhists realized that there is no abiding self, no unchanging "I" enthroned in our flesh. We are evanescent and ever changing. Even molecules disappear, upon inspection, into electrons, nucleons, quarks, and ever smaller tendencies. Whence this substantial delusion that greets us each morning? A fascinating question, to be sure, but the query should not validate the claim.

Meanwhile, our judgments and decisions originate from and are contained in a particular Weltanschauung (matrix of factual beliefs, value judgments, and ultimate goals; Dilthey), faith (orientation of the personality, to oneself, to one's neighbors, and to the universe; Smith), or system of memes (which respond to selection pressures independent of the carrier and their genes; Dawkins). But given the number of quarks in the universe and the number of synapses in our brains, our models must fall short of the reality with which we interact. The current "reality" of Western culture is militaristic capitalism. Missile systems receive far more funding than prenatal health care and education (let alone general education).

In the next few decades, a nanotech research and development lab will be within the means of many cultural groups who feel that their need to impose particular values (memes) warrants the violent abrogation of the life and freedom of others. Their options are frightening, including diamondoid tanks, designer viruses, and unimaginably thorough surveillance. Despite the inherent dangers, open development seems the only viable course of action.

The following logic seems inescapable:

1. Nanotechnology is coming and is potentially dangerous.

2. Successfully suppressing public development will merely relocate efforts into unaccountable arenas of the military.

3. Immense success in suppressing development in the United States, Europe, and Japan will only drive development into even less accountable regions of the planet.

In any event, if nanotechnology is successful in, say, providing us with the ability to construct a wide variety of arbitrary diamondoid structures with resource costs equivalent to pulp board, we can expect a great deal of such nanotechnologically machined materials in the biosphere.

In the context of an emerging nanotechnology, and with a historical awareness of the multiplicity of value systems that our species has entertained, the question is, what Weltanschauung or hybrid of memetic structures can lead us not only to survival but to beauty and delight?

All life, to date, supports and consists of an intricate molecular dance centered on the double helix of DNA. Around and within this dance swarm the molecular multitudes that make up our world. That these festivities gives rise to ant colonies and peacocks, computers and geodes, and such charming delusions as "you" and "me," is both bizarre and wonderful. Perhaps molecules deserve our reverence and deep gratitude. Are molecules sacred? Is anything?

BC Crandall, 8 November 1991

☐ Chapter 1 Molecular Engineering

The epigraphs are from Frank E. Harris, "Molecules," *Encyclopedia of Physics,* Second Edition, Rita Lerner and George Trigg, eds. (New York: VCH Publishers, Inc., 1991), 761, and Samuel C. Florman, *The Existential Pleasures of Engineering* (New York: St. Martin's Press, 1976), x. Florman concludes, "My proposition is that the nature of engineering has been misconceived. Analysis, rationality, materialism, and practical creativity do not preclude emotional fulfillment. They do not 'reduce' experience, as is so often claimed; they expand it. Engineering is superficial only to those who view it superficially. At the heart of engineering lies existential joy." 101.

1. One of the best visceral introductions to the scale of things is a film by Charles and Ray Eames, *Powers of Ten* (Santa Monica, California: Pyramid Film & Video, 1989), and the associated book, Philip and Phylis Morrison and the Office of Charles and Ray Eames, *Powers of Ten: About the Relative Size of Things in the Universe* (New York: Scientific American Books, 1982). This is a wonderful film that could serve as an excellent hypermedia front-end for the study of physical science.

2. The meter is "the fundamental unit of length (equivalent to 39.37 inches) in the metric system. It was defined in 1790 as one ten-millionth (10^{-7}) of the earth's quadrant passing through Paris, but was redefined in 1960 as the length equal to 1,650,763.73 wavelengths in a vacuum of the orange-red radiation of krypton 86." William Morris, ed., *The American Heritage Dictionary of the English Language* (Boston: Houghton Mifflin, 1969), 825.

3. Kenneth W. Ford, "The Large and the Small," in Timothy Ferris, ed., *The World Treasury of Physics, Astronomy, and Mathematics* (Boston: Little, Brown and Company, 1991),

22. First published in Kenneth W. Ford, *The World of Elementary Particles* (Cambridge: Cambridge University Press, 1958).

4. "For an office worker in London [the rate of atmospheric diffusion] means he's surrounded by and inhaling oxygen atoms that had been in Paris a few days before, breathed out by men smoking Gauloises, or emerging from temporary storage in microscopic cavities in upholstery surfaces or wall paint there. In a year's time, the oxygen atoms will have traveled a distance greater than the earth's circumference, and will be coming and going from every part of the globe. This is where things get interesting. Oxygen atoms last a very long time. . . . About one-sixth of the oxygen atoms you breathe in are released in the next breath, quite untouched by anything that went on inside. The same thing happens with everyone else who is breathing—your spouse, your boss, your friends, enemies and others—and because such terrific numbers of molecules are involved, because the stuff is so light and easily dispersed, it's statistically near certain that within a year's time you'll be breathing some of the oxygen, some of the exact same molecules, that they sucked in a year before." David Bodanis, *The Secret House: 24 Hours in the Strange and Unexpected World in Which We Spend our Nights and Days* (New York: Simon and Schuster, 1986), 215–216. This book is a splendid and highly entertaining discussion of the real but invisible molecular world that we all inhabit.

5. R. W. Schoenlein, L. A. Peteanu, R. A. Mathies, and C. V. Shank, "The First Step in Vision: Femtosecond Isomerization of Rhodopsin," *Science* 254 (18 October 1991): 412–415. "These measurements demonstrate that the first step in vision, the 11-*cis* → 11-*trans* torsional isomerization of the rhodopsin chromophore [retinal], is essentially complete in only 200 femtoseconds."

6. "Brownian motion in aqueous solution makes protein-sized parts shift by their own diameter roughly 10^6 times a second, they turn by a radian roughly 10^7 times per second, and they shift by an atomic diameter 10^{10} times a second. Thus they thoroughly and rapidly explore their environment and possible arrangements, becoming trapped when details of surface complementarity conspire to provide a deep energy well. The result can be rapid, reliable self-assembly of complex structures." K. Eric Drexler, "Strategies for Molecular Systems Engineering," in BC Crandall and James Lewis, eds., *Nanotechnology: Research and Perspectives* (Cambridge: The MIT Press, 1992), 117.

7. Kenneth W. Ford, "The Large and the Small," in Timothy Ferris, ed., *The World Treasury of Physics, Astronomy, and Mathematics* (Boston: Little, Brown and Company, 1991), 22. First published in Kenneth W. Ford, *The World of Elementary Particles* (Cambridge: Cambridge University Press, 1958).

8. "The death of a large star is a sudden and violent event. The star evolves peacefully for millions of years, passing through various stages of development, but when it runs out of nuclear fuel, it collapses under its own weight in less than a second. The most important events in the collapse are over in milliseconds. What follows is a supernova, a prodigious explosion more powerful than any since the big bang with which the universe began." Hans A. Bethe and Gerald Brown, "How a Super Nova Explodes," in Timothy Ferris, ed., *The World Treasury of Physics, Astronomy, and Mathematics* (Boston: Little, Brown and Company, 1991), 277.

9. Frank O. Copley, trans., *Lucretius, The Nature of Things* (New York: W. W. Norton & Company, 1977), xi. Lucretius (96?–55 BCE), a Roman poet, wrote *De Rerum Natura* (The Nature of Things) in the first century BCE based on the philosophy of Epicurus (342–270 BCE), which in turn developed from the work of Democritus.

10. Andrew van Melsen, "Atomism," *The Encyclopedia of Philosophy*, vol. 1 (New York: Macmillan Publishing Company, 1967), 196.

11. John Dalton, *A New System of Chemical Philosophy*, 1808. Reprinted in Louise B. Young, ed., *The Mystery of Matter* (New York: Oxford University Press, 1965), 34.

12. For a detailed exploration of a "triangular and tetrahedral system [that] uses 60-degree coordination instead of 90-degree coordination," see R. Buckminster Fuller, in collaboration with E. J. Applewhite, *Synergetics: Explorations in the Geometry of Thinking* (New York: Macmillan Publishing, 1975).

13. Dmitri Mendeléev, "The Periodic Law of the Chemical Elements," delivered before the Fellows of the Chemical Society, 1889. Reprinted in Louise B. Young, ed., *The Mystery of Matter* (New York: Oxford University Press, 1965), 46–49.

14. P. W. Atkins has written two splendid introductory texts to the molecular world that I cannot recommend too highly: *Molecules* (New York: W. H. Freeman and Company, 1987), and *Atoms, Electrons, and Change* (New York: W. H. Freeman and Company, 1991).

15. P. W. Atkins, *Molecules*, 22.

16. P. W. Atkins, *Molecules*, 37–38.

17. "In theory, all one needs to know in order to fold a protein into its biologically active shape is the sequence of its constituent amino acids. Why has nobody been able to put theory into practice?" Frederic M. Richards, "The Protein Folding Problem," *Scientific American* (January 1991): 54–63.

18. Lubert Stryer introduces his text on biochemistry announcing that "The genome is now an open book—any passage can be read. The cloning and sequencing of millions of base pairs of DNA have greatly enriched our understanding of genes and proteins. Indeed, recombinant technology has led to the integration of molecular genetics and protein chemistry. The intricate interplay of genotype and phenotype is now being unraveled at the molecular level. . . . The molecular circuitry of growth and development is coming into view." Lubert Stryer, *Biochemistry*, third edition (New York: W. H. Freeman and Company, 1988), xxv.

See also James Darnell, Harvey Lodish, and David Baltimore, *Molecular Cell Biology*, second edition (New York: W. H. Freeman and Company, 1990), vii. "We asserted in the preface to the first edition of this book [1986] that the reductionist approach and the new techniques of molecular biology would soon unify all experimental biology. Now, four years later, perhaps the only surprise is the speed and completeness with which biologists from fields formerly considered distant have embraced the new experimental approaches."

19. P. W. Atkins, *Atoms, Electrons, and Change*, 94.

20. "The rubber glove, with its red exterior and green interior, when stripped inside-outingly from off the left hand as red, now fits the right hand as green. First the left hand was conceptual and the right hand was nonconceptual—then the process of stripping off inside-outingly created the right hand. And then vice versa as the next strip-off occurs. Strip it off the right hand and there it is left again." R. Buckminster Fuller, in collaboration with E. J. Applewhite, *Synergetics: Explorations in the Geometry of Thinking* (New York: Macmillan Publishing, 1975), 232.

21. "Artificial life . . . is the computer scientist's Great Work as surely as the building of the Notre Dame cathedral on the Ile de France was the Great Work of the medieval artisan." Rudy Rucker, in Steven Levy, *Artificial Life: The Quest for a New Creation* (New York: Pantheon Books, 1992), 47.

22. All derivations from William Morris, ed., *The American Heritage Dictionary of the English Language* (Boston: Houghton Mifflin, 1969).

23. See Robert Serber, *The Los Alamos Primer: The First Lectures on How to Build An Atomic Bomb*, annotated by Robert Serber, edited by Richard Rhodes (Berkeley: University of California Press, 1992). First published as a "Limited Secret" Army report with an introductory preface which read, "The following notes are based on a set of five lectures given

by R. Serber during the first two weeks of April 1943, as an 'indoctrination course' in connection with the starting of the Los Alamos Project. The notes were written up by E. U. Condon." In the 1992 introduction, Rhodes writes, "The device tested at Trinity site in the New Mexico desert at 5:30 A.M. on July 16, 1945, was an implosion mechanism with a plutonium core. It exploded with a force equivalent to 18,600 tons of TNT, the first full-scale nuclear explosion on earth. I. I. Rabi watched it from a base camp some ten miles away: 'We were lying there, very tense, in the early dawn, and there were just a few streaks of gold in the east; you could see your neighbor very dimly. Those ten seconds [of the countdown] were the longest ten seconds that I ever experienced. Suddenly, there was an enormous flash of light, the brightest light I have ever seen or that I think anyone has ever seen. It blasted; it pounced; it bored its way right through you. It was a vision that was seen with more than the eye. It was seen to last forever. You would wish it would stop; altogether it lasted about two seconds. Finally it was over, diminishing, and we looked toward the place where the bomb had been; there was an enormous ball of fire which grew and grew and it rolled as it grew; it went up into the air, in yellow flashes and into scarlet and green. It looked menacing. It seemed to come towards one.... A new thing had just been born; a new control; a new understanding of man, which man had acquired over nature.'" xvii. While nuclear power and molecular engineering are distinct in many ways, they are similar in that both enable human activity at the scale of billions. A nuclear explosion is so powerful because it heats the fissioning material, in less than a millionth of a second, to about 10 billion degrees centigrade, temperatures seen hitherto only in stars.

24. Erwin Schrödinger, *What is Life?* (Oxford: Oxford University Press, 1944), reprinted in Louise B. Young, ed., *The Mystery of Matter* (New York: Oxford University Press, 1965), 433–460. See also a report on a recent meeting held to commemorate the fiftieth anniversary of Schrödinger's original series of lectures: Luke O'Neill, Michael Murphy, Richard B. Gallagher, "What Are We? Where Did We Come From? Where Are We Going?" *Science* 263 (14 January 1994): 181–182. These writers conclude, "We do not have much time left to prove that we are not the products of a lethal mutation."

25. Erwin Schrödinger, *What is Life?* (Oxford: Oxford University Press, 1944), quoted in James R. Newman, "Commentary on Erwin Schrödinger," James R. Newman, ed., *The World of Mathematics,* vol. 2 (New York: Simon and Schuster, 1956), 973.

26. James R. Newman, "Commentary on Erwin Schrödinger," in James R. Newman, ed., *The World of Mathematics,* vol. 2 (New York: Simon and Schuster, 1956), 973.

27. Steve J. Heims, *John von Neumann and Norbert Wiener: From Mathematics to the Technologies of Life and Death* (Cambridge: The MIT Press, 1980), 212. "His theory of self-reproducing automata is among von Neumann's most original work, and although he had elaborate plans for developing the theory in various directions, he became sidetracked by more practical projects, from computers to hydrogen weapons." See also John von Neumann, "Probabilistic Logics and the Synthesis of Reliable Organisms from Unreliable Components," in A. H. Taub, ed., *Collected Works,* vol. 5 (New York: Macmillan, 1963), 329–378; and John von Neumann, *Theory of Self-Reproducing Automata,* edited by A. W. Burks (Urbana: Illinois University Press, 1966).

28. Richard Feynman, "There's Plenty of Room at the Bottom," in BC Crandall and James Lewis, eds., *Nanotechnology: Research and Perspectives* (Cambridge: The MIT Press, 1992), 360.

29. Gerd Binnig, Heinrich Rohrer, C. Gerber, and E. Weibel, "Surface studies by scanning tunneling microscopy," *Physical Review Letters* 49 (1982): 57–61.

30. Ivan Amato, "Scanning Probe Microscopes Look Into New Territories," *Science* 261 (8 October 1993): 178.

31. For a detailed introduction to the mechanisms and instrumentation of STMs and

AFMs, see C. Julian Chen, *Introduction to Scanning Tunneling Microscopy* (New York: Oxford University Press, 1993).

32.   K. Eric Drexler, "Molecular engineering: An approach to the development of general capabilities for molecular manipulation," *Proceedings of the National Academy of Science, USA* 78:9 (September 1981): 5275–5278. In this article, Drexler also broaches the issue of cryonic preservation of the brain after "death," the subsequent reconstruction of neural patterns within the frozen tissue, and thus the regeneration of memories in an anticipated nanotechnological rebirth.

33.   K. Eric Drexler, "When molecules will do the work," *Smithsonian* (November 1982).

34.   "Although hopes for 'buckyballs' border on the messianic, practical uses for the talented molecules have yet to emerge. They may find their true calling in 21st century microelectronics and nanotechnology." Hugh Aldersey-Williams, "The Third Coming of Carbon," *Technology Review* (January 1994): 54–62.

35.   K. Eric Drexler, *Engines of Creation: Challenges and Choices of the Last Technological Revolution* (Garden City, New York: Anchor Press/Doubleday, 1986). In his foreword, Marvin Minsky writes, "*Engines of Creation* sets us on the threshold of genuinely significant changes; nanotechnology could have more effect on our material existence than those last two great inventions in that domain—the replacement of sticks and stones by metals and cements and the harnessing of electricity. Similarly, we can compare the possible effects of artificial intelligence on how we think—and on how we might come to think about ourselves—with only two earlier inventions: those of language and of writing." vii.

36.   Christopher Langton, ed., *Artificial Life: The Proceedings of an Interdisciplinary Workshop in the Synthesis and Simulation of Living Systems Held September, 1987 in Los Alamos, New Mexico* (Redwood City, California: Addison-Wesley Publishing Company, 1989), xv.

37.   Conrad Schneiker, "NanoTechnology with Feynman Machines: Scanning Tunneling Engineering and Artificial Life," in Christopher Langton, ed., *Artificial Life* (Redwood City, California: Addison-Wesley Publishing Company, 1989), 444.

38.   Richard Feldmann, "Applying engineering principles to the design of a cellular biology" (paper presented at the celebration of the 100th anniversary of Bayer Pharmacia at Boppard, Germany, 6 October 1988).

39.   Hans Moravec, *Mind Children: The Future of Robot and Human Intelligence* (Cambridge: Harvard University Press, 1988), 73.

40.   BC Crandall and James Lewis, eds., *Nanotechnology: Research and Perspectives* (Cambridge: The MIT Press, 1992).

41.   Philip Ball, "Small Problems," *Nature* 362 (11 March 1993): 123.

42.   G. M. Shedd and P. E. Russell, "The scanning tunneling microscope as a tool for nanofabrication," *Nanotechnology* 1 (July 1990).

43.   Richard Stengel, "Best of '90: Science and Technology," *Time* (31 December 1990), 51. "Smallest Advertisement. Using a powerful microscope, IBM researchers lined up individual xenon atoms to spell out the company's initials. That clever display of know-how got magnified pictures of the minuscule logo into newspapers all over the world—for free." Including this issue of *Time*.

44.   J. A. Armstrong, "New Frontiers in Computing and Telecommunications," *Creativity!* (an IBM in-house magazine), June 1991, 1–6.

45.   D. Swinbanks, "Japan Will Fund Major Nanotechnology Project," *Nature* (1991): 650.

46.   *Science* 254 (29 November 1991): 1300–1341.

47.   Andrew Pollack, "Atom by Atom, Scientists Build 'Invisible' Machines of the Future," *New York Times*, 26 November 1991, C1–C3.

48. Michael Flynn, *The Nanotech Chronicles* (New York: Baen Books, 1991). "The genie is dangerous . . . but he's already out of the bottle. We daren't go forward, and we cannot go back. The nanny will be built, but we mustn't build it. No way out. No solution. There had to be a solution. His soul rebelled against the notion that there may not always be solutions." 145.

49. K. Eric Drexler, *Nanosystems: Molecular Machinery, Manufacturing, and Computing* (New York: John Wiley & Sons, Inc., 1992).

50. *Nanosystems* was praised by William Goddard, Director of the Materials and Molecular Simulation Center at the California Institute of Technology: "With this book, Drexler has established the field of molecular nanotechnology. The detailed analyses show quantum chemists and synthetic chemists how to build upon their knowledge of bonds and molecules to develop the manufacturing systems of nanotechnology, and show physicists and engineers how to scale down their concepts of macroscopic systems to the level of molecules." The book was also endorsed by Marvin Minsky, professor of computer science at MIT: "Devices enormously smaller than before will remodel engineering, chemistry, medicine, and computer science. How can we understand machines so small? *Nanosystems* covers it all: power and strength, friction and wear, thermal noise and quantum uncertainty. This is the book for starting the next century of engineering." Both comments, *Nanosystems,* i.

Philip Ball, reviewing *Nanosystems* for *Nature,* was more critical. "Drexler's vision of the nanoscale future is so limited, consisting solely of diamond-based ('diamondoid') machines performing every task by mechanical means, he places nanotechnology in a straight-jacket that prevents the ends from the current means converging. . . . In effect, Drexler is proposing to invent a new chemistry, which he calls 'machine-phase chemistry'. But the question is whether this challenge is worth taking up, rather than building on the achievements so far. . . . There is next to no mention of fullerenes, perhaps the ideal nanoscale building blocks, and none of Sumio Iijima's carbon nanotubes, surely the closest we have yet come to nano girders." Philip Ball, "Small Problems," *Nature* 362 (11 March 1993): 123.

For a complete account of Ball's perspective, see his recent book: *Philip Ball, Designing the Molecular World: Chemistry at the Frontier* (Princeton: Princeton University Press, 1994). Ball describes the generation of amphiphilic-bilayer membranes and their formation into natural cells and artificial "protocells" in chapters 7 and 8.

51. See K. Eric Drexler, *Nanosystems: Molecular Machinery, Manufacturing, and Computing* (New York: John Wiley & Sons, Inc., 1992), 1.

52. See National Science Foundation Report, *NSF Blue Ribbon Panel on High Performance Computing—From Desktop to Teraflop: Exploiting the U.S. Lead in High Performance Computing,* by Lewis Branscomb, (Chairman), Theodore Belytschko, Peter Bridenbaugh, Teresa Chay, Jeff Dozier, Gary S. Grest, Edward F. Hayes, Barry Honig, Neal Lane, William A. Lester Jr., Gregory J. McRae, James A. Sethian, Burton Smith, Mary Vernon. (National Science Foundation publication no. NSF 93-205, 19 October 1993). "Justin Rattner of Intel estimated that in 1996 microprocessors with clock speeds of 200 MHz may power an 800 Mflops peak speed workstation. . . . [Rattner and others] held out the likelihood that in 1997 microprocessors may be available at 1 gigaflop; a desktop PC might be available with this speed for $10,000 or less. . . . Today one can purchase a mid-range workstation with a clock speed of 200 MHz for an entry price of $40,000 to $50,000." 14.

53. Philip Ball and Laura Garwin, "Science at the atomic scale," *Nature* (27 February 1992): 761.

54. Christopher Langton, ed., *Artificial Life II* (Redwood City, California: Addison-Wesley Publishing Company, 1992).

55. M. Mitchell Waldrop *Complexity: The Emerging Science at the Edge of Order and Chaos* (New York: Simon & Schuster, 1992), and Steven Levy, *Artificial Life: The Quest for a New Creation* (New York: Pantheon Books, 1992).

56. J. Doyne Farmer and Alletta Belin, "Artificial Life: The Coming Evolution," in Christopher Langton, ed., *Artificial Life II* (Redwood City, California: Addison-Wesley Publishing Company, 1992), 815.

57. "New Journals Reflect Growing Interest in Euro, Eco, Nano, and Neuro," *Science Watch* (October 1992); reprint, *The Scientist* (31 May 1993): 14.

58. Lynn Simarski, "National Science Board Approves Creation of National Nanofabrication User's Network," National Science Foundation Press Release (NSF-PR 93–88) 22 November 1993.

59. Ivan Amato, "Scanning Probe Microscopes Look Into New Territories" *Science* 262 (8 October 1993): 178.

60. Neil Gross, Emily Smith, and John Carey, "Windows on the World of Atoms," *Business Week* (30 August 1993), 62–64. "Roughly five SPM makers have sprung up since 1986. Last year, the largest, Digital Instruments in Santa Barbara, California, sold $21 million worth."

61. G. Nunes Jr. and M. R. Freeman, "Picosecond Resolution in Scanning Tunneling Microscopy," *Science* 262 (12 November 1993): 1029.

62. "Shape of things to come: Molecular STM," *Science News* (26 June 1993): 407, reporting the work of Vickie M. Hallmark and Shirley Chiang published in *Physical Review Letters* (14 June 1993).

63. M. F. Crommie, C. P. Lutz, and D. M. Eigler, "Confinement of Electrons to Quantum Corrals on a Metal Surface," *Science* 262 (8 October 1993): 218.

64. I. Peterson, "Wrapping carbon into superstrong tubes," *Science News* (3 April 1993): 214.

65. Hugh Aldersey-Williams, "The Third Coming of Carbon," *Technology Review* (January 1994): 61.

66. "Tubules self-assemble smaller than DNA," *Science News* (21 August 1993): 122, reporting the work of Akira Harada published in *Nature* (5 August 1993).

67. R. Lipkin, "New nanotubes self-assemble on command," *Science News* (27 November 1993): 357, reporting the work of M. Reza Ghadiri published in *Nature* (25 November 1993).

68. Joel M. Schnur, "Lipid Tubules: A Paradigm for Molecularly Engineered Structures," *Science* (10 December 1993): 1669.

69. Michelle Hoffman, "Motor Molecules on the Move," *Science* 256 (26 June 1992): 1758.

70. John Travis, "Innovative Techniques on Display at Boston Meeting: Stepping Out With Kinesin," *Science* (27 August 1993): 1112–1113.

71. Robert R. Birge, "Protein-Based Optical Computing and Memories," *IEEE Computer* (November 1992): 56–67. For an accessible introduction to bacteriorhodopsin-based computing, see Robert R. Birge, "Protein-Based Computers," *Scientific American* 272:3 (March 1995): 90.

72. Ivan Amato, "Harvesting Light With Molecular Leaves," *Science* 261 (10 September 1993): 1388.

73. "The logic to molecular computers," *Science News* (24 July 1993): 63, reporting the work of A. Prasanna de Silva, H. Q. Nimal Guaratne, and Colin P. McCoy, "A molecular photoionic AND gate based on fluorescent signalling," published in *Nature* 364 (1 July 1993): 42–44.

74. Marvin L. Cohen, "Predicting Useful Materials," *Science* 261 (16 July 1993): 307. See also Chunming Niu, Yuan Z. Lu, and Charles M. Lieber, "Experimental Realization of the Covalent Solid Carbon Nitride," *Science* 261 (16 July 1993): 334.

75. Karen F. Schmidt, "Evolution in a Test Tube: Harnessing Darwin's Theory to Design New Molecules," *Science News* (7 August 1993): 90–93.

76. Stephanie Forrest, "Genetic Algorithms: Principles of Natural Selection Applied to Computation," *Science* 261 (13 August 1993): 872.

77. Stephanie Forrest, "Genetic Algorithms: Principles of Natural Selection Applied to Computation," *Science* 261 (13 August 1993): 878. See, for example, a recent ad for Cheyenne's InocuLAN™, which promises "Real-time virus protection . . . [monitoring] your entire enterprise while in the background, trapping viruses before they can spread." *Network Computing* (15 January 1994): 23.

78. John H. Holland, "Complex Adaptive Systems," *Dædalus* (Winter 1992): 17.

79. "Cellular automata are abstract dynamical systems that play in discrete mathematics a role comparable to that played in the mathematics of the continuum by partial differential equations. In terms of structure as well as applications, they are the computer scientist's counterpart to the physicist's concept of a 'field' governed by 'field equations.' It is not surprising that they have been reinvented innumerable times under different names and within different disciplines. The canonical attribution is to Ulam and von Neumann (ca. 1950).

"In cellular automata, space is represented by a uniform array. To each site of the array, or *cell* (whence the name 'cellular'), there is associated a state variable ranging over a finite set—typically just a few bits worth of data. Time advances in discrete steps, and the dynamics is given by an explicit rule—say a look up table—through which at every step each cell determines its new state from the current state of its neighbors." Tommaso Toffoli, "Cellular Automata," *Encyclopedia of Physics*, Second Edition, Rita Lerner and George Trigg, eds. (New York: VCH Publishers, Inc., 1991), 126.

80. James A. Reggia, Steven L. Armentrout, Hui-Hsien Chou, and Yun Peng, "Simple Systems that Exhibit Self-Directed Replication," *Science* 259 (26 February 1993): 1282.

81. Christopher Langton, *Physica D* 10 (1984): 135.

82. See Ivan Amato, "Capturing Chemical Evolution in a Jar," *Science* 255 (14 February 1992): 800. Amato reports on the work of Julius Rebek, Jr., an organic chemist at MIT, who built a simple, two-part molecule that, once formed, acts as a catalytic template for the construction of similar molecules from the two original components.

83. James A. Reggia, Steven L. Armentrout, Hui-Hsien Chou, and Yun Peng, "Simple Systems that Exhibit Self-Directed Replication," *Science* 259 (26 February 1993): 1287.

84. Current-architecture computers must dissipate at least $kT \ln 2$ of energy (about 3×10^{-21} joules at room temperature) for each bit of information they erase or otherwise throw away.

85. Charles H. Bennett, "Logical Reversibility of Computation," *IBM Journal Research Devices* 17 (1973): p. 525–532; reprint in Harvey S. Leff and Andrew F. Rex, eds., *Maxwell's Demon: Entropy, Information, Computing* (Princeton, New Jersey: Princeton University Press, 1990), 197–204. See also Charles H. Bennett, "Notes on the History of Reversible Computing, *IBM Journal Research Devices* 32 (1988): 16–23, reprinted in Leff and Rex, 281–288; Ralph C. Merkle, "Reversible Electronic Logic Using Switches," *Nanotechnology* 4 (1993): 21–40; and Ralph C. Merkle, "Two Types of Mechanical Reversible Logic, *Nanotechnology* 4 (1993): 114–131.

86. Faye Flam, "Researchers Defy the Physical Limits of Computation: Rolling Back the Cost of Computing," *Science* 260 (16 April 1993): 290–291.

87. Faye Flam, "Researchers Defy the Physical Limits of Computation: Chemists Simulate Smart Beakers," *Science* 260 (16 April 1993): 291. See also A. Hjelmfelt, F. W. Schneider, and J. Ross, "Pattern Recognition in Coupled Kinetic Systems," *Science* 260 (16 April 1993): 335.

88. "Artificial neural networks are models of highly parallel and adaptive computation, based loosely on current theories of brain structure and activity. There are many different neural net architectures and algorithms, but the basic algorithm which all artificial neural nets share is the same. Assume a collection of simple processors ('units') and a high degree of connectivity, each connection having a *weight* associated with it. Each unit computes a weighted sum of its inputs (then may plug this sum into some nonlinear function), assumes a new level of activation, and sends an output signal to the units to which it is connected. In many of the models, the network settles into a stable global state under the influence of external input that represents an interpretation of the input or a function computed on the input. This settling process, called relaxation, performs a parallel search." Evan W. Steeg, "Neural Networks, Adaptive Optimization, and RNA Secondary Structure Prediction," in Lawrence Hunter, ed., *Artificial Intelligence and Molecular Biology* (Menlo Park, CA and Cambridge, MA: AAAI Press/The MIT Press, 1993), 136–137.

89. Jean-Pierre Banâtre and Daniel Le Métayer, "Programming by Multiset Transformation," *Communications of the ACM* 36 (January 1993): 98–111.

90. Jack LeTourneau, "Implementing Simple Finite Multisets with Natural Numbers," (Prime Arithmetics, Inc., 3410 Geary Boulevard, Suite 311, San Francisco, CA 94118).

91. "A program in general consists of data and code that operates on the data. Object-oriented programming encapsulates in an object some data and programs to operate on the data: the data is the state of the object, and the code is the behavior of the object." Won Kim, *Introduction to Object-Oriented Databases* (Cambridge: The MIT Press, 1990), 9.

92. Russell Ruthen, "Flat Chemistry: Enormous Polymer Sheets Promise Unusual Properties," *Scientific American* (April 1993): 26. See also Edwin L. Thomas, "Gigamolecules in Flatland," *Science* 259 (1 January 1993): 43–44; and S. I. Stupp, S. Son, H. C. Lin, and L. S. Li, "Synthesis of Two-Dimensional Polymers," *Science* 259 (1 January 1993): 59–63.

93. Harold J. Morowitz, *Beginnings of Cellular Life: Metabolism Recapitulates Biogenesis* (New Haven: Yale University Press, 1992), 9, 178.

☐ Chapter 2   In-Vivo Nanoscope and the "Two-Week Revolution"

The epigraphs are from Francis Bacon, *The Works of Francis Bacon,* vol. 4 (London: Longman & Co., 1960), 47, and Alfred North Whitehead, *Science and the Modern World* (New York: New American Library, 1963), 107, both quoted in Don Ihde, *Instrumental Realism: The Interface between Philosophy of Science and Philosophy of Technology* (Bloomington: Indiana University Press, 1991), 62 and 67, respectively.

1. Since 1983, the National Oceanic and Atmospheric Administration (NOAA) has managed a small fleet of orbiting satellites, the Landsat robots, that generate multispectral images of the earth.

2. The secrets of the thymus gland are being unlocked by researchers as you read this. This particular story will be understood well before the invention of the in-vivo nanoscope. However, there will be no shortage of other equally compelling cellular mysteries that will propel the development of nanotechnology-based instrumentation.

3. K. Eric Drexler, "Molecular Machinery and Molecular Electronic Devices," *Molecular Electronic Devices II* (New York: Marcel Dekker, 1987).

4. The idea that a great wave of change would occur during the first two weeks after an assembler is built originated in the summer of 1980. Roger Gregory, Mark S. Miller, and Rowland King thought of this while brainstorming at a meeting in King of Prussia, Pennsylvania.

5. Morton Grosser, *Gossamer Odyssey* (New York: Houghton Mifflin, 1981).

6. Stephane Groueff, *Manhattan Project: The Untold Story of the Making of the Atomic Bomb* (Boston: Little, Brown and Company, 1967), 218.

7. See Carver Mead and Lynn Conway, *Introduction to VLSI Systems* (Redwood City, California: Addison-Wesley, 1980), 47.

☐ **Chapter 3   Cosmetic Nanosurgery**

1. See Dr. Seuss, "The Sneetches," *The Sneetches and Other Stories* (New York: Random House, 1961) 21. In the story of Sneetches there were two kinds of creatures. Star-Bellied Sneetches had stars plain to see, but the others had none and were deprived of much glee. Then a Fix-it-Up Chappie drove on to their beaches with tools that could change all the stars on those Sneetches:

All the rest of that day, on those wild screaming beaches,
The Fix-it-Up Chappie kept fixing up Sneetches.
Off again! On again!
In again! Out again!
Through the machines they raced round and about again,
Changing their stars every minute or two.
They kept paying money. They kept running through
Until neither the Plain nor the Star-Bellies knew
Whether this one was that one . . . or that one was this one
Or which one was what one . . . or what one was who.

2. See Gregory Fahy, "Possible Medical Applications of Nanotechnology: Hints from the Field of Aging Research," BC Crandall and James Lewis, eds., *Nanotechnology: Research and Perspectives* (Cambridge: The MIT Press, 1992), 251–267.

3. See Edward Reifman, "Diamond Teeth" (this volume, chap. 4).

4. A surrogate "human" skin has recently been developed for large-area skin grafts. It protects the underlying tissues and promotes scarless recovery, but it is far from the exact match to human skin that would be required for cosmetic testing.

☐ **Chapter 5   Early Applications**

The epigraph is from Pat Cadigan, "Pretty Boy Crossover," in *The Year's Best Science Fiction: Fourth Annual Collection,* ed. Gardner Dozois (New York: St. Martin's Press, 1987), quoted in Scott Bukatman, *Terminal Identity: The Virtual Subject in Postmodern Science Fiction* (Durham, NC: Duke University Press, 1993), 256.

1. See, for example, *The Diamond Age; or, A Young Lady's Illustrated Primer* (New York: Bantam Books, 1995). A nanomachined book figures prominently in this clever science fiction tale.

2. See Robert Reich, *The Work of Nations: Preparing Ourselves for 21st Century Capitalism* (New York: Alfred A. Knopf, 1991) 208. "Regardless of how your job is officially classi-

fied (manufacturing, service, managerial, technical, secretarial, and so on), or the industry in which you work (automotive, steel, computer, advertising, finance, food processing), your real competitive position in the world economy is coming to depend on the function you perform in it. . . . The fortunes of routine producers are declining. In-person servers are also becoming poorer, although their fates are less clear-cut. But symbolic analysts [aka knowledge workers]—who solve, identify, and broker new problems—are, by and large, succeeding in the world economy."

3. See, for example, Vernor Vinge, "Technological Singularity," *Whole Earth Review* (Winter 1993): 88–95. "The acceleration of technological progress has been the central feature of this century. We are on the edge of change comparable to the rise of human life on Earth. The precise cause of this change is the imminent creation by technology of entities with greater-than-human intelligence. Science may achieve this breakthrough by several means. . . . Three [of these] possibilities depend on improvements in computer hardware. Progress in hardware has followed an amazingly steady curve in the last few decades. Based on this trend, I believe that the creation of greater-than-human intelligence will occur during the next thirty years. . . . [To] be more specific: I'll be surprised if this event occurs before 2005 or after 2030. . . . This change [the singularity] will be a throwing-away of all human rules, perhaps in the blink of an eye—an exponential runaway beyond any hope of control." 89. See also Vinge's fictional tales about the singularity: *Across Realtime* (New York: Baen Books, 1991). See also Hans Moravec, *Mind Children: The Future of Robot and Human Intelligence* (Cambridge: Harvard University Press, 1988). "The postbiological world will host a great range of individuals constituted from libraries of accumulated knowledge. In its early stages, as it evolves from the world we know, the scale and function of these individuals will be approximately that of humans. But this transitional stage will be just a launching point for a rapid evolution in many novel directions . . ." 125.

4. See J. Storrs Hall, "Utility Fog" (this volume, chap. 10) for an alternate approach to building generalized intelligent stuff.

5. R. Buckminster Fuller, in collaboration with E. J. Applewhite, *Synergetics: Explorations in the Geometry of Thinking* (New York: Macmillan Publishing Co., Inc., 1975). See also Robert Curl and Richard Smalley, "Fullerenes," *Scientific American* (October 1991): 54–63.

6. K. Eric Drexler, *Nanosystems: Molecular Machinery, Manufacturing, and Computing* (New York: John Wiley & Sons, Inc., 1992), 371. "A CPU-scale system containing 10^6 transistor-like interlocks (constructed [as described]) can fit within a 400 nm cube. . . . The power consumption for a 1 GHz, CPU-scale system is estimated to be ~60 nW, performing $> 10^{16}$ instructions per second per watt."

☐ **Chapter 6   The Companion: A Very Personal Computer**

The epigraphs are from Michael Heim, "The Erotic Ontology of Cyberspace," Michael Benedikt, ed., *Cyberspace: First Steps* (Cambridge: The MIT Press, 1991), 60; and William Gibson, "Academy Leader," Michael Benedikt, ed., *Cyberspace: First Steps* (Cambridge: The MIT Press, 1991), 29.

1. K. Eric Drexler, *Engines of Creation: Challenges and Choices of the Last Technological Revolution,* Garden City, New York: Anchor Press/Doubleday, 1986.

2. Data storage estimates are based on the following figures:

- 10,000 hours of movies and video: 108×10^{12} bytes; 10,000 hours of music: 2.7×10^{12} bytes; 10,000,000 books: 100×10^{12} bytes; 10,000 maps: 13×10^9 bytes; courseware and knowledge bases: 1×10^{12} bytes. Total: 221.7×10^{12} bytes.

- Size of a video frame: 2000×2000 pixels $\times 3$ bytes (24 bit color) = 12,000,000 bytes.

At fifty frames per second, and 200:1 compression, 10.8×10^9 bytes per hour (for nonstereo-scopic material) or 16.2×10^9 bytes per 90-minute movie.

- Music: "A 60-minute musical selection recorded in stereo with PCM [pulse code modulation] with a sampling rate of 44.1KHz and 16-bit quantization generates over 5 billion bits." Ken Polhman, "The Compact Disc Formats," *Journal of the Audio Engineering Society* (April 1988): 250–280.
- Maps: 10,000 2000 $\times$ 2000 pixel $\times$ 24 bit images, each with one megabyte overhead for object description (buildings, roads, etc.): 12 Mb + 1 Mb = 13 Mb $\times$ 10,000 = 130 Gb, or, with 10:1 compression, 13 Gb.

3. J. R. Asker, "Motorola Proposes 77 Lightsats for Global Mobile Phone Service," *Aviation Week & Space Technology* (2 July 1990).

4. Michael Alexander, "Eye-tracking system may let eyes replace the mouse," *Computerworld* (16 September 1991).

5. John Gantz, "GIS meets GPS," *Computer Graphics World* (October 1990).

6. See Ashok Bindra, "English/Spanish translator works in real time," *Electronic Engineering Times* (27 April 1992), and Jannis Moutafis, "Automatic Telephone Translator," *Electronic Engineering Times* (16 March 1992).

7. Today, portable phones are often given away as premiums and prizes; phone companies make a tidy profit on the service charges for their use.

8. See John G. Maguire, "Interactive Magazines," *Computer Graphics World* (August 1991).

9. Software "agents," computer programs that independently carry out specialized tasks, can be helpful here.

10. See Nicholas P. Negroponte, "Products and Services for Computer Networks," *Scientific American* (September 1991).

11. See Misha A. Mahowald and Carver Mead, "The Silicon Retina," *Scientific American* (May 1991).

Chapter 7   Trivial (Uses of) Nanotechnology

Portions of this chapter previously appeared in *Cryonics,* the magazine of the Alcor Life Extension Foundation, Inc., Riverside, California.

1. Ancestral eyes may have started as reflector surfaces, concentrating light on sensitive nerve patches. Lenses seem to have come along later. The optic nerves attenuate light a bit and cause a blind spot where they plunge through the light receptor surface. In contrast, octopus eyes are wired the sensible way: the nerves all come out the back side of the retina. See Richard Dawkins, *The Blind Watchmaker* (New York: W. W. Norton, 1986), 93.

Chapter 8   Nanotech Hobbies

1. T. Koppel, "Learning How Bacteria Swim Could Set New Gears in Motion," *Scientific American* (September 1991).

Chapter 9   Phased Array Optics

1. I. P. Kaminow, *An Introduction to Electrooptic Devices* (New York: Academic Press, 1974).

2. F. J. Kahn, "Electronically Tunable Birefringence," U.S. Patent 3,694,053, assigned to Bell Telephone, 1972.

3. Jeffery Soreff, private communication, 1991.

4. For a preview of such a suit, see the film *Predator* (Hollywood, CA: Twentieth Century Fox Film Corporation, 1988).

5. Howard Rheingold, *Virtual Reality,* (New York: Summit Books/Simon & Schuster, 1991).

☐ Chapter 10   Utility Fog: The Stuff That Dreams Are Made Of

Portions of this chapter previously appeared in *Extropy: The Journal of Transhumanist Thought* 6, nos. 2 and 3 (1994).

The epigraphs are from William Shakespeare, *The Tempest,* Act IV, Scene I, *The Riverside Shakespeare* (Boston: Houghton Mifflin Company, 1974) 1630; Arthur C. Clarke, *Against the Fall of Night* (New York: Gnome Press, 1949); E. E. Smith, *Skylark Three* (New York: Pyramid, 1963).

1. See John Papiewski, "The Companion" (this volume, chap. 6) and Brian Wowk, "Phased Array Optics (chap. 9).

2. See B. A. Hubermann, ed., *The Ecology of Computation* (Cambridge: The MIT Press, 1985).

3. See R. Buckminster Fuller, in collaboration with E. J. Applewhite, *Synergetics: Explorations in the Geometry of Thinking* (New York: Macmillan Publishing Co., Inc., 1975). See also U.S. Patent 2,986,241, granted 30 May 1961, for the octet truss.

4. "The Reynolds number is the most important scaling parameter in aerodynamics and a controlling guideline for the overall behavior and analysis of a wide variety of fluid flows." Daniel Bershader, "Fluid Physics," in *Encyclopedia of Physics,* 2d ed., edited by Rita Lerner and George Trigg (New York: VCH Publishers, Inc., 1991), 408.

5. See K. Eric Drexler, *Nanosystems: Molecular Machinery, Manufacturing, and Computing* (New York: John Wiley & Sons, Inc., 1992).

6. Entropy cannot decrease, but erasing a bit would decrease entropy. Thus the machine erasing a bit must emit enough heat to raise the entropy of the rest of the universe by a corresponding amount. For one bit, the amount of energy is comparable to the thermal energy of motion of a single atom.

7. Ralph C. Merkle, "Reversible Electronic Logic Using Switches," *Nanotechnology* 4 (1993) 21–40, and "Two Types of Mechanical Reversible Logic," *Nanotechnology* 4 (1993) 114–131. See also J. Storrs Hall, "Nanocomputers and Reversible Logic," *Nanotechnology* 5 (1994).

The concluding epigraph is from William Shakespeare, *The Tempest,* Act IV, Scene I, *The Riverside Shakespeare* (Boston: Houghton Mifflin Company, 1974), 1630.

☐ Postscript

The epigraphs are from Richard Preston, *The Hot Zone* (New York: Random House, 1994) 72; and Hans Moravec, *Mind Children: The Future of Robot and Human Intelligence* (Cambridge: Harvard University Press, 1988), 129.

1. See Laurie Garrett, *The Coming Plague: Newly Emerging Diseases in a World Out of Balance* (New York: Farrar, Straus and Giroux, 1994). In her introduction, Garrett quotes Nobel laureate (Medicine, 1958) Joshua Lederberg, president of Rockefeller University, as

he opened a conference in 1989 to address the challenge of newly emerging microbes: "Nature isn't benign. The bottom lines: the units of natural selection—DNA, sometimes RNA elements—are by no means neatly packaged in discrete organisms. They all share the entire biosphere. The survival of the human species is *not* a preordained evolutionary program. Abundant sources of genetic variation exist for viruses to learn new tricks, not necessarily confined to what happens routinely, or even frequently." 6.

2.   Richard Preston, *The Hot Zone* (New York: Random House, 1994), 287.

3.   See Ben R. Finney and Eric M. Jones, eds., *Interstellar Migration and the Human Experience* (Berkeley: University of California Press, 1985).

The concluding epigraph is from Ludwig Wittgenstein, *Vermischte Bemerkungen* (Frankfurt, 1977) 14, quoted in Richard Rorty, *Philosophy and the Mirror of Nature* (Princeton: Princeton University Press, 1979), vii.

☐ Contributors

The epigraph is from Gilles Deleuze, *Foucault,* Seán Hand, trans. and ed. (Minneapolis: University of Minnesota Press, 1988), 104.

■ Index

I define an Index as a Sign determined by its Dynamic object by virtue of being in a real relation to it.
—C. S. Peirce

Entries followed by an f indicate a figure; n indicates a note.